# THE FONTS COACH

by Cheri Robinson
and Daniel Gray

New Riders Publishing
Carmel, Indiana

**The Fonts Coach**

By Cheri Robinson and Daniel Gray

Published by:
New Riders Publishing
11711 N. College Ave., Suite 140
Carmel, IN 46032 USA
1-800-835-3203



Printed in the United States of America 1 2 3 4 5 6 7 8 9 0

Library of Congress Cataloging-in-Publication Data is available.

**Publisher**
*David P. Ewing*

**Associate Publisher**
*Tim Huddleston*

**Marketing Manager**
*Brad Koch*

**Acquisitions Editor**
*John Pont*

**Managing Editor**
*Cheri Robinson*

**Product Director**
*Rob Tidrow*

**Production Editor**
*Lisa D. Wagner*

**Editors**
*JoAnna Arnott*
*Geneil Breeze*
*Rob Tidrow*

**Technical Editor**
*John Cornicello*

**Editorial Secretary**
*Karen Opal*

**Book Design and Production**
*Katy Bodenmiller*
*Christine Cook*
*Amy Peppler-Adams*
*Angela M. Pozdol*
*Michelle Self*
*Susan Sheperd*
*Greg Simsic*
*Alyssa Yesh*

**Proofreaders**
*Mark Enochs*
*Carla Hall-Batton*
*Howard Jones*
*John Kane*
*R. Sean Medlock*
*Tim Montgomery*
*Tammy Tidrow*
*Angie Trzepacz*
*Steve Weiss*

**Indexed by**
*John Sleeva*

# DEDICATION

A writer stares at a blank page for years before knowing what belongs on it. Without the support of family and friends, the page remains blank and is swept away by the winds of life.

Thanks Rich, Erin, and Blair.

# ABOUT THE AUTHORS

**Cheri Robinson** is Managing Editor at New Riders Publishing. She has 15 years of experience in the publishing field. During her tenure at New Riders, she has edited and developed a number of titles, including *Inside LANtastic*, *A Guide to CD-ROM*, and *CorelDRAW! on Command*. Before joining New Riders she was a Senior Editor for Que Corporation. Ms. Robinson also was employed as a graphic designer for several trade publications in the hardware industry.

**Daniel Gray** is a journeyman artist who has been involved in both traditional and electronic publishing for more than 10 years. Dan's electronic publishing experience includes stints at the drawing board, in the darkroom, in systems management, and with personal computers. Dan has worked on dozens of publications including *The Princeton Packet* and *Women's Wear Daily*. He founded and publishes *Danzai Wire*, America's only independent journal for Suzuki automobile owners.

Dan is currently a Graphic Systems Analyst for the Continental Corporation and uses both MS-DOS and Macintosh platforms to publish a wide range of periodicals, from simple newsletters to four-color magazines.

Readers are invited to contact Daniel Gray through CompuServe. His CIS number is 71210,667.

# ACKNOWLEDGMENTS

The authors thank the following individuals for their contributions to this book:

John Pont for finding the tech editor and ensuring the material arrived on time. Also for hunting down and contracting the authors of the programs included on the bonus disk.

Rob Tidrow for his helpful advice and insightful comments.

The editorial staff at New Riders Publishing, including Lisa Wagner, JoAnna Arnott, and Geneil Breeze.

John Cornicello for ensuring the technical accuracy of the book and for compiling the contents of *The Fonts Coach* bonus disk. John compiled the disk by using the DTPForum on CompuServe. He can be reached through CompuServe # 76702,1410.

The production and indexing staff of Prentice Hall Computer Publishing, for the expeditious handling of this book.

Rob Tidrow wishes to thank Jim O'Gara and Sheila Thomas of Altsys corporation for their help acquiring Fontographer for Windows.


# TRADEMARK ACKNOWLEDGMENTS

## WARNING AND DISCLAIMER

# CONTENTS AT A GLANCE

# TABLE OF CONTENTS

XI

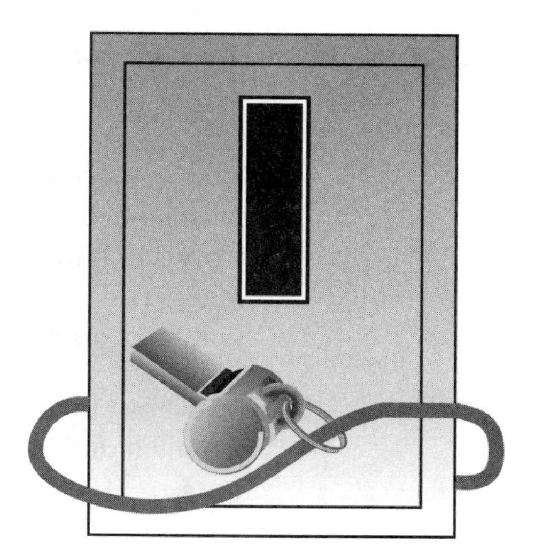


# INTRODUCTION

Picture this scenario: For the last week you have worked diligently on an important business proposal. You spent many late nights working away on your computer's word processing program. Your computer has saved you much time. You edited and spell-checked your work in a fraction of the time that those tasks would have taken just a few short years ago.

The word processing program also includes some other features that you vow to learn more about after you finish the proposal. You make a mental note to open up the documentation and learn more about using fonts, rules, and styles. You feel that you just don't have the time now to be bothered by learning something brand new.

The day of the presentation arrives. You present your proposal. It looks neat enough. You haven't misspelled any words, and the pages are all neatly typed and single spaced. Your boss nods his approval. You can't help wondering, however, what your competition is going to come up with.

After you present your proposal, your co-worker stands and passes out the counter-proposal. You look in amazement at the proposal. It looks as if it came from a professional desktop publisher! The text on the page appears in different styles of type. The headlines appear in a larger size than the body of the document.

Your boss takes a real interest in this proposal, shuffling your proposal underneath his notebook. Maybe it's time to learn about fonts.

## WELCOME TO THE PERSONAL TRAINER SERIES

This book is just one in a series of books that provides a common-sense, no-frills approach to shortening the learning curve involved in using today's technology. With this book, you learn step-by-step how to demystify the subject of fonts.

You won't find long, boring discussions on conceptual gibberish. On the contrary, this book contains practice sessions that give you a chance to use the knowledge that you gain from the brief and easily understood discussions. As you thumb through the pages, you will see notes from the coach. The coach is your personal trainer. He stands ready to answer your questions and to lead you through examples. You'll find simple tables and illustrations to guide you through the text.

Each book in the Personal Trainer Series comes equipped with a special disk, provided to help you get a head start on learning the subject area you've chosen. The disk in *The Fonts Coach*, for example, includes TrueType fonts for both Windows 3.1 and the Macintosh.

*The Fonts Coach* does not pretend to be the definitive guide to using fonts, but it does teach you the things you need to know to get the job done. From terminology to step-by-step instructions, you'll have a much better understanding of fonts after you finish reading this book.

## OTHER BOOKS IN THE SERIES

The Personal Trainer Series also includes the following books:

★ *The Graphics Coach*, by Katherine Murray

★ *The Modem Coach*, by Dana Blankenhorn

# WHO SHOULD USE THIS BOOK?

Just about anybody who uses a computer can take advantage of the information that this book has to offer. From the busy professional who must put together business reports to the person who is in charge of producing the newsletter for the elementary school, this book contains the information you need. Specifically, *The Fonts Coach* is for the following users:

★ People who have not used fonts, but are not new to computers (you may, like many people, have mastered the basic tasks of several software programs, but not previously learned the basics of fonts).

★ People who must learn basic graphic design quickly. Maybe you are responsible for setting up a newsletter for your organization.

★ New computer users who want to learn what all the fuss is about in using fonts.

★ People who have just begun to use Windows 3.1.

# HIGHLIGHTS OF *THE FONTS COACH*

*The Fonts Coach* is divided into three parts. Part I, "Font Basics," explains the basics of working with fonts. Part II, "Understanding Computer Font Technology," explains the differences among the various font formats, and Part III, "Working with Fonts," provides basic design guidelines to help you construct good-looking documents. Part IV, "Appendixes," contains a vendor list and a glossary. Specifically, *The Fonts Coach* contains the following chapters:

Chapter 1, "Why Use Fonts?," explains in simple terms what fonts can do for you. You learn about the personality that fonts project and how to use this personality to reinforce the message that you want to transmit to your readers. You also learn the various categories of font styles.

Chapter 2, "Understanding the Terminology," provides the historical background of the evolution of fonts. You learn how typography evolved and why the terminology that is used today relates to this centuries-old art. You also learn the parts that make up a letter and the technique you use to measure and mark up type.

Chapter 3, "Understanding Font Technology," briefly describes the different font formats. The advantages and disadvantages of the font types are discussed. You learn about bit-mapped fonts, scalable fonts, PostScript fonts, and TrueType fonts.

Chapter 4, "Understanding TrueType," explains how this font technology came about. You learn to install and remove TrueType fonts. This chapter also discusses printing and managing your TrueType fonts.

Chapter 5, "Understanding PostScript Type 1," thoroughly explains this font type. You learn proper installation and the problems that you can encounter with printing Type 1 fonts. You learn the benefits of using these fonts and also some of the disadvantages.

Chapter 6, “Shopping for Fonts,” shows you where you can go to buy (or even receive free) fonts of all types. This chapter discusses the major manufacturers, mail-order vendors, shareware, and freeware sources for fonts. You also learn what you need to know to find the correct fonts for you. You learn about quality considerations, cost considerations, and technical support considerations.

Chapter 7, “Before You Begin,” prepares you to successfully create professional-quality documents. You learn to break some of the habits you may have picked up over the years. You learn about the common typographical errors that new users make and how to avoid making these mistakes. This chapter also discusses using special keyboard characters.

Chapter 8, “Putting It All Together,” is a basic design guide of do’s and don’ts to help you create good-looking documents. You learn how to construct a page and the basic elements that go into building a page. You learn about font alignment, size, and spacing. This chapter also discusses copyfitting considerations.

Chapter 9, “Manipulating Fonts,” discusses font editors. You learn the benefits of using a font editor and how you should choose the font editor that is best for you. This chapter also discusses developing special effects for fonts. You learn how you can rotate or enhance the fonts that you have.

Appendix A is a vendor list that is complete with addresses and phones numbers of companies that provide fonts or font services.

Appendix B contains a glossary of typographical and font terms. The definitions are brief and easy to understand.

Now that you know the overall game plan for the book, you can kick off your learning session with fonts.

# THE FONTS COACH BONUS DISK

Included with this book is *The Fonts Coach* bonus disk, which includes a number of shareware and freeware programs that you can use with your Windows and Macintosh systems. Included on the disk is a copy of PKUNZIP.EXE, which will help you uncompress the files on the disk if you are using Windows.

First copy the files(s) that you want to use onto your hard drive, start DOS session, and type **PKUNZIP *filename*** to "unzip" the file (if it has a .ZIP extension). You need to substitute the name of the file for ***filename***. Look for a file called README.TXT or a similar name to see what special procedures you need to do to install and use the program.

If the program or font is shareware, please be sure to read the licensing agreement and, if you decide to use the program or font, please pay the licensing fee to the author of the program or font. The software printer font, Video Terminal Screen, for example, is provided as shareware by the following author and requires a $10.00 licensing fee if you decide to use the font:

E.A. BEHL Technical Publishing Services
2663 Red Oak Court
Clearwater, Florida U.S.A. 34621-2319

# PART I

## FONT BASICS

# WHY USE FONTS?

You may be wondering why a whole book should be devoted to the subject of fonts. If you own a Macintosh or IBM-compatible computer, you have access to many ways of printing letters, numbers, and other types of characters. During an average day, however, you read or sort through an avalanche of printed material. From the morning newspaper, to your business correspondence, to the menu from which you order your lunch, to the advertisements you receive in the mail, and finally the magazine you read before you go to bed, you come in contact with a wide assortment of printed words that appear in a wide assortment of sizes and fonts.

In this chapter, you learn what fonts can do for you. Specifically, you cover the following points:

**GAME PLAN**

- ☐ Understanding the tone of fonts
- ☐ Understanding the differences among the four general font categories

*continues*

*continued*

- Understanding when and why you should use each type of font
- Examining the four subcategories of fonts

# THE TONE OF FONTS

The fonts used in the makeup of documents can play as large a role as the actual words used to broadcast the message. *Fonts*, which are complete sets of letters, numbers, and punctuation marks that have a consistent typeface, influence the appearance of a printed document more so than any other visual element. They set the mood. Why not bring this same power to the documents you create?

Some fonts evoke a serious, no-nonsense tone; others conjure up a more whimsical feeling. You learn to understand the differences among the four general categories of fonts and when and why you should use each type. These four categories include text, display, decorative, and specialty. This chapter goes a step further by introducing you to the subcategories of the four general categories of fonts. After you understand the differences between the categories, you will have no trouble creating attention-getting documents that project the tone you want to convey.

# EXAMINING WHAT FONTS CAN DO FOR YOU

As you thumb through the stack of mail you receive each day, have you ever wondered why a few of the advertisements catch your eye? Although you may stop and glance at those advertisements that feature bright colors and glossy photographs, a few of the pieces that catch your attention may contain no color or flashy

graphics. The understated elegance of type can do much to enhance the impact of a printed document.

Before computerized desktop publishing, the average computer user had few options when it came to designing documents and choosing fonts. Now, however, a whole new world has opened up. Today's sophisticated word processing and desktop publishing software and the development of TrueType fonts have brought graphic design down to the level of the most basic user. No longer is typography the arena of just high-priced designers. No matter if you are a busy executive creating a financial report, a secretary setting up a business form, or a Girl Scout leader creating a newsletter, you can learn and use the power of fonts.

## REACHING THE READER

No mistaking it; you live in a visual world. From the time you wake up until the time you go to bed, media of all types compete for your attention. In days past, newspapers, books, and even magazines were published with little emphasis placed on readability. Publishers could count on people sitting down to read their publications.

Today, however, the world is a much different place. People are busy. They have more items to distract them. Television sets, VCRs, video games, and computers all serve to distract and entertain the average person. Today's publications have responded to this phenomena. Newspapers such as *USA Today* and magazines such as *Time* employ flashy graphics and dynamic typestyles to entice people to read the articles.

How does this apply to you? You must keep your readers in mind. Invite them to read your text. If you want your information read, you must make your document compelling. Professional, well-designed documents can make the difference.

## The Coach Says...

Target your reader. Ask yourself who will read your document and what the purpose is for the document. If you are writing a letter to a friend, the font you choose can be casual and friendly. A business letter, however, must project a more serious tone.

Figure 1.1 illustrates the difference that applying a few well-chosen fonts can make in the visual appeal of a document. Both of these documents contain the same information. The document on the left is printed entirely in Courier, which is a font that resembles standard typewriter print. The document on the right uses professional looking fonts.

**Figure 1.1:**

*Viewing the difference that fonts make.*

**AAA HOME PRODUCTS, INC. PRESS RELEASE**

1338 Shadow Lakes Drive North
Carmel, Indiana 46032

For Immediate Release

March 17, 1993

NEW HOUSEWORK GIZMO INTRODUCED

The Research and Development staff of AAA Home Products, Inc. has introduced an amazing new gizmo that is sure to become a staple in all American homes. Today's busy homemakers don't have time to spend long hours scrubbing and waxing floors.

With the new, Scrubby Waxy, they no longer need to tediously do this tiresome work. The revolutionary Scrubby Waxy enables the overworked homemaker to fill one side of the sponge mop with Scrubby detergent and the other side with Waxy floor wax. No more messy mops and buckets of water.

The Scrubby detergent leaves no detergent residue, so no rinsing is required. And best of all, you can apply the Waxy wax right after mopping. No waiting for the floor to dry.

So throw away that old mop and bucket, and use the new Scrubby Waxy system. You'll have more time to do the things that you enjoy.

For more information, contact Richard Robinson at (317) 848-7541.

Figure 1.1:

Continued

AAA Home Products, Inc.
1338 Shadow Lakes Drive North
Carmel, Indiana 46032

***PRESS RELEASE***

March 17, 1993

**New Time-saving Gizmo Introduced**

The Research and Development staff of AAA Home Products, Inc. has introduced an amazing new gizmo that is sure to become a staple in all American homes. Today's busy homemakers don't have time to spend long hours scrubbing and waxing floors.

With the new, Scrubby Waxy, they no longer need to tediously do this tiresome work. The revolutionary Scrubby Waxy enables the overworked homemaker to fill one side of the sponge mop with Scrubby detergent and the other side with Waxy floor wax. No more messy mops and buckets of water.

The Scrubby detergent leaves no detergent residue, so no rinsing is required. And best of all, you can apply the Waxy wax right after mopping. No waiting for the floor to dry.

So throw away that old mop and bucket, and use the new Scrubby Waxy system. You'll have more time to do the things that you enjoy.

For more information, contact Richard Robinson at (317) 848-7541.

Although the document on the left clearly states all the information, notice how using just a few fonts can really make the difference. The headline on the document on the right quickly draws the reader in. The font used on this press release is called Lucida Bright. It was chosen because it is highly readable and also photocopies well. This document is definitely very professional looking, but doesn't require a whole lot of design skill to create.

## FLEXIBILITY IN TONE

Fonts have many personalities. You can find a font that conveys almost any message or feeling that you want to express. Fonts can

express feelings such as happiness, sadness, tranquility, anger, and playfulness. You can find casual fonts or ornate fonts. Fonts can make you think of an event, such as a wedding, or of a particular time in history, such as the old west.

Reading the tone of a font is much like reading the feelings of a person by studying facial features and body language. You probably are adept at studying a person's face and posture to determine how you in turn should react to them. Rigid posture with arms crossed, for example, conveys anger. Furrowed brows with fingers touching the forehead conveys a person deep in thought. Graphic designers have long used the same type of technique to bring emphasis to the printed word. You can use these same principles to choose fonts that will emphasize the message you want to broadcast.

Exercise 1.1 demonstrates this philosophy. The column on the left contains the word "type" in various fonts. Each of these fonts conveys a message. The column on the right contains descriptive words. Just for fun, try matching a word in the left column with a word from the right column.

You can take the information you learn in Exercise 1.1 one step further by placing the descriptive word from the right column and assigning it the corresponding font from the left column (see fig. 1.2). The words have instant impact.

## ACCURACY IN PRESENTATION

Just as choosing the right font can be rewarding, choosing the wrong one can be disasterous. As you look through the lists of fonts from which you can choose, immediately mark off any fonts that are obviously wrong for the job. If you want to create a business memo, for example, you should not choose an ornate script typeface. You will want to select a more professional looking font.

## Exercise 1.1
## Understanding the Tone of Type

| | |
|---|---|
| TYPE | Western |
| Type | Historical |
| Type | Scary |
| Type | Blueprint |
| Type | Fancy |
| Type | Winter |
| Type | Casual |
| TYPE | Sixties |
| TYPE | Childish |
| type | Bamboo |
| tYpe | Neon |

Figures 1.3 and 1.4 illustrate two versions of a wedding invitation. The version shown in figure 1.3 uses a font named PostCrypt. The version shown in figure 1.4 uses a traditional font to which most people would be accustomed to seeing. The font used in figure 1.3 is defintely inappropriate, unless of course the bride is marrying Dracula! The font used in figure 1.4 accurately conveys the message of the special occasion.

The preceding example was pretty outrageous, but you should keep it in mind as you choose your fonts.

**Figure 1.2:**

Font selection can add impact to the message.

SCARY
Blueprint
Fancy
Western
Historical
Casual
Sixties
WINTER
BAMBOO
neon
CHiLDiSH

**Figure 1.3:**

One way to get your point across.

MR. AND MRS. RICHARD ROBINSON
REQUEST THE HONOUR OF YOUR PRESENCE
AT THE MARRIAGE OF THEIR DAUGHTER
ERIN MARIE ROBINSON
TO
MICHAEL PATRICK COLLINS
ON JULY 7, 1993
1:30 P.M. AT
OUR LADY OF MOUNT CARMEL CHURCH
CARMEL, INDIANA

*Mr. and Mrs. Richard Robinson*
*request the honour of your presence*
*at the marriage of their daughter*
*Erin Marie Robinson*
*to*
*Michael Patrick Collins*
*on July 7, 1993*
*1:30 p.m. at*
*Our Lady of Mount Carmel Church*
*Carmel, Indiana*

**Figure 1.4:** A more appropriate way to get your point across.

**The Coach Says...**

Remember the preceding examples as you choose the fonts for your documents. Just use common sense. Use a strong font to set the mood and then choose a font that enables the reader to clearly read and understand your message.

Now that you have a better understanding of the power that fonts can bring to your printed documents, you can examine the categories of fonts and the subcategories.

## UNDERSTANDING FONT CATEGORIES

Thumbing through a font specimen book may intimidate you. You have hundreds of choices you can make, and you wonder which fonts you should choose. You will feel much more

comfortable, however, when you realize that all fonts can be divided into four general categories. You use each category for a specific job or effect. These font categories include:

- ★ Text
- ★ Display
- ★ Decorative
- ★ Specialty

The following sections discuss the four general categories. You learn the definition of each category, when to use the category, and typical fonts that make up the category.

## TEXT FONTS

Text fonts are the primary building blocks of a good type library. You use text fonts in long passages of text in which you need a high degree of legibility. *Legibility* is the quality of the font that enables the reader to quickly read and understand your message.

Text fonts make up the body copy, or the regular reading matter of your document. The body copy generally is smaller than the display type used for headlines. Text fonts usually measure between 86 points and 14 points in size.

> **The Coach Says...**
> Don't worry that you don't understand the terminology now. If you open any magazine, newspaper, or book, the words that make up the body of the articles are printed in text fonts. You learn the terminology and the method of measuring type in Chapter 2.

Just as that conservative grey suit of yours takes you to almost any occasion, a good conservative text font can serve you equally as well. If you are just beginning to use fonts, it is better to err on the side of simplicity than to go overboard and purchase too many fonts. You can always purchase more fonts as you become more adept at discerning the differences between them.

**The Coach Says...**

If you are interested in more information about obtaining fonts, turn to Chapter 6. This chapter provides information on what to look for when you select fonts and where to go to purchase them.

You probably have an appropriate text font already. Windows 3.1, for example, comes with a basic set of fonts. Within this basic set, Times New Roman is a good basic text font. The Macintosh also comes with a version of Times Roman (see fig. 1.5).

Times Roman

**Figure 1.5:** The Times New Roman text font.

Other common text fonts include Garamond, Palatino, New Century Schoolbook, and Bookman. Figure 1.6 illustrates these common text faces. Each of these fonts has subtle differences, but all have one thing in common: they have a high degree of readability.

**Figure 1.6:**

Examples of text fonts.

Garamond

Palatino

New Century Schoolbook

Bookman

## DISPLAY FONTS

Display fonts add interest and visual impact to your documents. You most often use display fonts for headlines, signs, or advertisements. Unlike text fonts, display fonts don't need to have a high degree of readability. Headlines and signs usually contain just a few lines of text; therefore, readability is not affected.

Because headlines usually are short attention getters, display fonts can have much more personality than text faces. Typically, display fonts employ stylistic extremes. Display fonts can contain wide variations in the thickness of letters; they might feature exaggerated rounded or elongated characters. Other display faces just offer strong, bold characters. All display faces serve to draw the reader to your messages.

Figure 1.7 illustrates several display fonts. Poster Bodoni is an example of a font that contains variations in the thickness of letters. Other examples of display fonts include Avant Garde, which features large round and stylized letters, and Helvetica Bold, which features strong, bold characters.

Poster Bodoni
Eurostyle Bold
Revue
STENCIL

**Figure 1.7:** Examples of display fonts.

**The Coach Says...**
Although you can use some of your text fonts as display fonts, display fonts rarely can be used as text fonts. Display fonts are not created to be read easily at smaller point sizes.

## DECORATIVE FONTS

Decorative fonts add punch to your documents and are fun to use. You can think of decorative fonts as you do formal evening attire. You bring out your tuxedo or black evening dress for special events. You don't bring them out for every occasion. The same can be said for decorative fonts.

Decorative fonts, when used sparingly, can set the mood that you want to create. Decorative fonts belong on party invitations, advertisements, posters, and signs. Don't use decorative fonts on business reports or other no-nonsense documents. Decorative fonts convey "fun." Figure 1.8 shows some examples of decorative fonts.

**Figure 1.8:**

Decorative fonts.

Ballet Engraved
Casper Open Face
Dear teacher

Note that you can incorporate decorative fonts into company logos. Because these fonts convey an instant message, you often can create instant name recognition for your company. Consider Coca-Cola, for example. Their logo is instantly recognizable because the font is so distinctive.

## SPECIALTY FONTS

Specialty fonts, or pi fonts as they are sometimes called, are those fonts that may not contain any letters at all. These fonts include all types of symbols and characters such as arrows, checkmarks, pencils, or other small characters. Other specialty fonts may include fractions, small capital letters, or special borders. By using specialty fonts, you can really jazz up your documents.

The most common use of specialty fonts is in a scientific or mathematical environment. If you have ever tried to type complicated mathematical equations or scientific formulas, you can appreciate how these fonts simplify the task. Figure 1.9 illustrates a few specialty fonts.

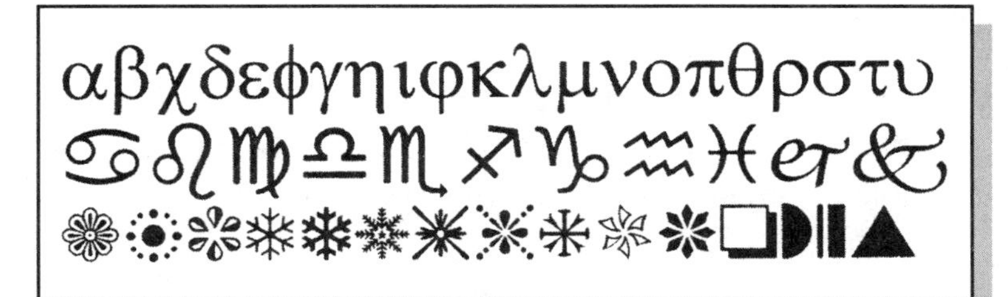

**Figure 1.9:**

Specialty fonts.

# SUBCATEGORIES OF FONTS

The preceding section discussed the four general categories of fonts. Fonts can be broken down further into subcategories. These subcategories consist of six distinctive designs.

By learning to tell the difference between these designs, you can learn to identify fonts. If you can narrow your search down to one category, you can probably find the font you are looking for in a type specimen book. The subcategories are as follows:

- ★ Uncial
- ★ Old style
- ★ Transitional
- ★ Modern
- ★ Egyptian
- ★ Script

Each of these categories marks a distinct stage in font history. Do not be confused into thinking that the fonts you use today are from the modern category. The fonts from which you can choose can come from any of the categories. These labels simply mark the changes that took place in the font evolutionary process. This evolution spans over three centuries. New fonts that are developed today can be assigned to one of the six subcategories.

## UNCIAL

Uncial (pronounced un-shial) fonts are the oldest classification of fonts. The word *uncial* comes from the Latin word uncus, which means crooked. These fonts resemble the hand lettering that monks used in the Middle Ages.

Uncial fonts are a decorative font and should be used when you want the look of hand lettering. You can successfully use uncial fonts for certificates, awards, and invitations.

## OLD STYLE, TRANSITIONAL, AND MODERN

Old Style, Transitional, and Modern fonts often are considered as one unit of type. Old Style fonts originated in 1617. Transitional fonts originated in 1757 and Modern fonts were developed in 1788. Although these three types are distinctive, they are often considered as one unit because all three types include Roman letters.

Roman fonts include the vast majority of text fonts. The main characteristic of Roman fonts are the serifs. *Serifs* are the tiny strokes at the end of the top and bottom of letters. The serifs guide the reader's eyes across the page, making these fonts have a high degree of readability. The text of this book is printed in a serif font.

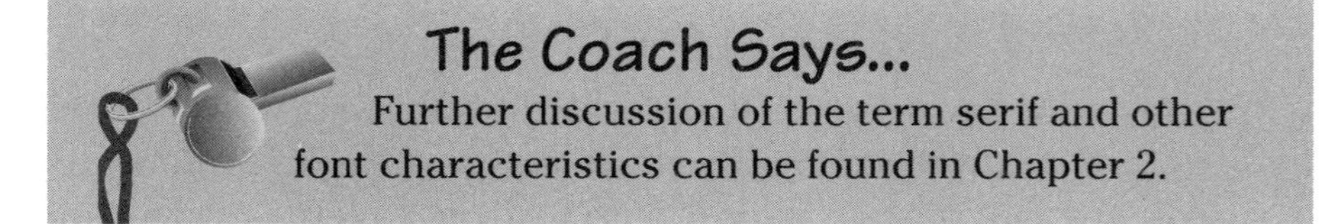

The Coach Says...
Further discussion of the term serif and other font characteristics can be found in Chapter 2.


Figure 1.10 illustrates an example of each Roman style of type. Notice that the serifs of the Old Style typeface are more curved than the serifs of the Transitional or Modern faces. The letters of

the Old Style font have an even stroke, whereas the letters of the Transitional and Modern faces show much more contrast in weight.

Garamond

New Baskerville

Bodoni

**Figure 1.10:** Examples of Old Style, Transitional, and Modern fonts.

**The Coach Says...**

Don't be discouraged if you cannot tell the difference between these font categories. You must train your eye to look for the fine distinctions in the weight of the letters and the serifs. Most type specimen books list the category to which a font belongs.

## EGYPTIAN

Egyptian fonts evolved after Napoleon returned from Egypt. Designers of these fonts borrowed characteristics from the previous font styles to create a distinctive look. Egyptian fonts include letters that are made up of strong and thick line weight. The letters have heavy slab serifs.

You can use Egyptian fonts as text fonts, but because they have such a strong look, they make excellent display fonts.

## SCRIPT

Script fonts are the easiest to identify. They imitate handwritten letters. Script fonts are perfect for short invitations or certificates. Never use Script fonts as text fonts for long documents because they are too difficult to read. Also, never use all caps when you are typing a document in Script fonts.

Table 1.1 provides examples of each type of font subcategory.

**Table 1.1**
**Subcategories of Fonts**

| *Subcategory* | *Font Examples* |
|---|---|
| Uncial | Black Chancery<br>Caligula<br>Lucida Blackletter |
| Old Style | Garamond Bold<br>Lucida Bright<br>Revive |
| Transitional | Book Antiqua<br>Century Schoolbook<br>Imperium |
| Script | Civiltype<br>Freeport<br>Lucida Calligraphy |

## INSTANT REPLAY

In this chapter, you learned the following:

INSTANT REPLAY

- ✔ Understanding the tone of fonts
- ✔ Understanding the differences among the four general font categories
- ✔ Understanding when and why you should use each type of font
- ✔ Examining the four subcategories of fonts

# UNDERSTANDING THE TERMINOLOGY

Now that you are excited about using fonts, you should learn some of the terminology used in this field. Although you can study many years and never achieve the status of a true typographer, learning some of the basics can help you better use fonts as you design your documents.

**GAME PLAN**

- ☐ Understanding the history of typography
- ☐ Understanding the anatomy of a letter
- ☐ Understanding the picas and points measurement system

## WHAT IS TYPOGRAPHY?

Remember when you first used your computer? The terminology seemed strange. You had to learn such terms as *operating system, windows, folder,* and *disk drive*. Using fonts successfully also employs special terms. This section introduces you to the basic language of typography. The words may seem peculiar at first, but soon you will be able to impress the professionals at your local quick-print shop with your typographic vocabulary.

*Typography* is defined as the study of the style and arrangement of printed words. This definition is overly simplistic, but basically it is correct. The types of fonts that you choose, the size of the font you use, and the way in which you arrange the letters on the page all make up the art of typography.

Understanding typography begins by understanding the development of the written alphabet. Letters are the symbols that convey your message. Through the centuries, mankind has used many types of symbols to communicate. The following section discusses the history of typography from the earliest stages into the modern day.

## A BRIEF HISTORY LESSON

Don't panic. This history lesson is short, and you won't be tested on the material. Understanding the terminology, however, separates the novice from the professional. Skim through this chapter now, and then mark it with a bookmark so that you can turn to it as a reference.

Man always has had a need to record events or keep track of important information. Although some civilizations relied on verbal communication, written language by far preserves a lasting record. From the earliest cave drawings, to Egyptian hieroglyphics, to the eventual development of the traditional alphabet, written symbols have communicated thoughts, events, and concepts.

## PICTURE WRITING

The first forms of written communication were pictographs. *Pictographs* are simple drawings of everyday objects, such as animals, food, and weapons. Archaeologists have found this early form of communication on the walls of the dwellings of the cavemen. These drawings often were of the animals the cavemen hunted or the weapons that they used.

> **The Coach Says...**
>
> If you are a Windows or Macintosh user, you should be familiar with pictographs. The icons that you see on your desktop are pictographs that represent programs or utilities. Figure 2.1 shows two common icons. The trash can icon should be familiar to Macintosh users. You use the trash can to discard files that you no longer need. The Windows File Manager icon represents the organizational structure of your computer's hard disk.

**Figure 2.1:** Computer pictographs.

As people wanted to communicate more abstract thoughts, however, another type of symbol was required. Man began to combine more than one pictograph to communicate such thoughts as happiness and danger. Symbols such as these are called *ideographs*. An example of a modern day ideograph is the familiar skull and crossbones symbol. Even children know that this symbol represents danger (see fig. 2.2).

**Figure 2.2:**

The familiar skull and crossbones ideograph.

> **The Coach Says...**
>
> Today, pictographs and ideographs are still in use. Communication is still visual much of the time. Good examples of modern-day visual communicators are the international symbols used on road signs.

This form of picture writing lasted for many centuries. Although this type of writing served the purpose, it did have several disadvantages. One disadvantage was that the symbols were complex to draw and the variations of symbols were many. Ideograph writing also was difficult to learn and required tedious drawing.

## ALPHABET SOUP

As civilizations became more sophisticated and people began to travel and trade goods with each other, a simplified writing system was needed so that merchants could keep business ledgers. Even during ancient days, accountants were sticklers for balanced ledgers.

The Phoenicians were the first to develop a revolutionary written communication. They created a form of writing that used symbols to represent sounds in speech rather than pictures that stood for objects or ideas.

They used a different symbol for each spoken sound of their language. Most of the symbols evolved from the original pictograph system. The pictograph symbol for the word ox (aleph), for example, was modified to stand for the sound "A." This writing system greatly reduced the time it took to record business transactions.

Around 1000 B.C., the Greeks began to adopt the Phoenician alphabet. As a civilization of great philosophers, they saw the potential to use the system to record and preserve knowledge. The Greeks used the Phoenician symbols, but added their own language. *Aleph*, therefore, became *alpha*.

The Phoenician alphabet contained no vowels. The Phoenicians did not need vowels because their system of writing was more like a business shorthand. The Greeks, however, needed a more complete alphabet to record their great literary works. They added five vowels.

**The Coach Says...**

The word alphabet is derived from two Greek words: *alpha* and *beta*.

The Romans further refined the alphabet. They modified eight letters and added the letters F and Q. At first the Roman alphabet contained 23 letters. After further refinement, the Romans added the letters U, W, and J to their alphabet. The modern day alphabet was born. The Romans also dropped the Greek names for letters, such as alpha, beta, gamma, for the simpler A, B, C. Figure 2.3 illustrates the differences between the Phoenician, Greek, and Roman alphabets.

**Figure 2.3:**

The early alphabets.

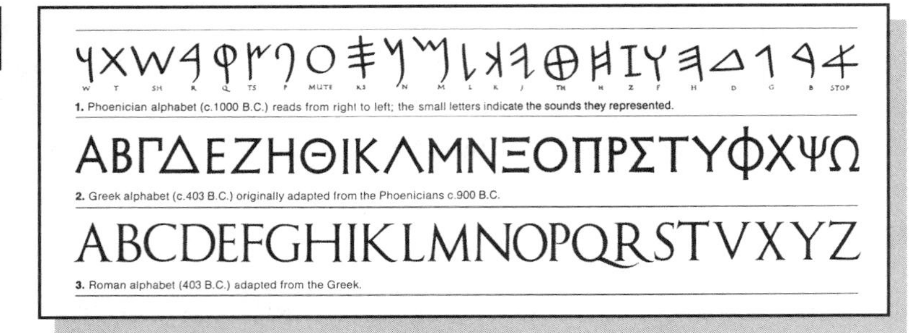


## THE DEVELOPMENT OF MOVABLE TYPE

The Phoenician, Greek, and Roman alphabets contained only uppercase letters. During the Middle Ages, however, monks developed lowercase, or *miniscule* letters. They painstakingly wrote books on parchment. Only the most wealthy and educated could own books. As the desire for knowledge grew, a method was needed to produce books in a more efficient manner.

Johannes Gutenberg, a German goldsmith, set out to develop a method of mass producing books. He took his knowledge of metal stamping and applied it to printing. He knew how to cast objects in metal and how to stamp images or letters on the objects. His goal was to take this skill and use it in the field of printing.

This task was no easy matter. The writing typical of the mid-fifteenth century consisted of heavy black letters that required uppercase, lowercase, combinations of letters known as *ligatures*, and punctuation marks. The font that Gutenberg developed contained nearly 300 characters!

### The Coach Says...

Ligatures are joined pairs of letters. The shapes of some letters when placed together cause awkward spacing problems. Consider the letter

combination fi. Because the letter "f" contains a projecting top stroke and the letter "i" contains an ascending dot, the two letters can appear to be crowded on top of each other. Typing this combination as a ligature solves this problem (see fig. 2.4).

æ Æ

**Figure 2.4:**

An example of a ligature.

### The Coach Says...

In early writing, the punctuations marks that you know today were not used. Words were separated by dots or slashes. It wasn't until the fifteenth century that traditional punctuation marks were instituted.

Gutenberg cast each individual character into lead. Each character was a raised reverse symbol so that the character appeared correctly after it was printed on paper. This process was much like the traditional stamp and stamp pad sets of today. The raised letter looked backwards at first glance, but printed correctly once it was stamped on the paper. The raised letters were mirror images of the actual printed letter.

Figure 2.5 shows a line of casted type. The letters were strung together to form words and sentences. The process of stringing

the letters together to form words and sentences was called *casting*. The printer had to actually cast each letter onto a wooden form. This form of printing was much faster than the hand-lettering approach of the monks, and the books also could be mass produced.

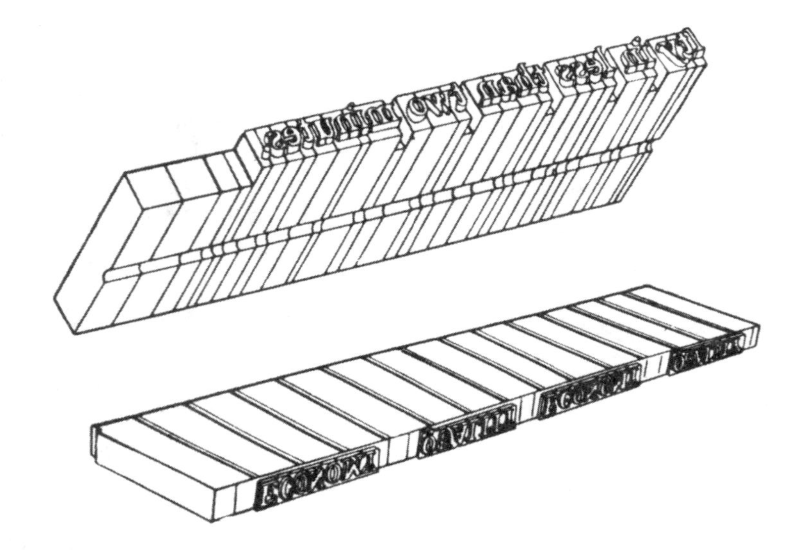

**Figure 2.5:** A piece of type.

**The Coach Says...**

The earliest surviving printed book of Gutenberg's is the Bible. The book contained 1,286 pages and was issued in two volumes. Historians estimate that between 180 and 200 copies were printed. Of this number, only 21 copies are intact.

## THE DEVELOPMENT OF MACHINE CASTING

By the late 1800s, machines began to do the letter casting that before this time had been done by hand. Gutenberg set each line

of type, printed the job, and then carefully cleaned and reused the letters. The development of machines enabled operators to type on keyboards similar to today's typewriters. As the typist struck the keys, molds of the letters filled with molten lead. The lead would solidify to produce lines of type. These machines were known as Linotype machines (see fig. 2.6).

**Figure 2.6:** A Linotype machine.

The machine-set type was cast and printed and then melted down to be recast. This method was much faster and more cost efficient than hand setting type. The speed at which type could be set was limited only to the typist's ability. This method of creating type lasted until this century.

## MODERN DAY TYPESETTING

During the late 1960s, typesetting took a giant step toward the computer-based society of today with the development of phototypesetting. With this method, an operator types at a keyboard. As each letter is typed, a high-intensity light shines through a negative of the letter and projects the image of the letter onto photosensitive film (see fig. 2.7). The film is then developed in a processor in much the same manner as the film in your camera is developed.

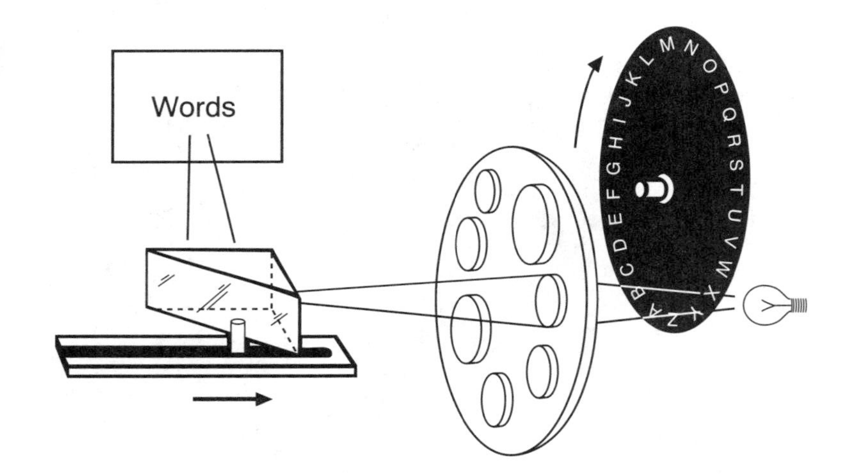


**Figure 2.7:** Understanding the phototypesetting process.

### The Coach Says...

With phototypesetting, the letters of a particular style of type are stored on a circular film. The circular film contains the letters (fonts) of a specific typeface. As the operator types the letters, the circular film spins to the appropriate letter, and the beam of light shines through the impression and projects the letter onto film.

The art of typesetting became available to the average computer user when Apple developed the Macintosh computer in the early '80s. For the first time, fonts were available in typical word processing software. The capacity to use fonts spread even further after Microsoft introduced Windows 3.1. With the development of TrueType fonts, an inexpensive and versatile type library became available to even the most inexperienced computer users.

**The Coach Says...**

For more information on the development of computerized fonts, see Chapter 3. For specific information on TrueType fonts, see Chapter 5.

You now have a good background in the history of typography. Hopefully, the lesson wasn't too painful. Before you can successfully use fonts, you should have an understanding of the anatomy of individual letters. The next section teaches you to recognize these characteristics.

# DISSECTING A LETTER

Each letter has specific components. These components include the strokes, dots, and lines that come together to make up the letter. A careful study of these components can help you tell the difference between styles of type.

Individual letters provide many clues to properly classifying a typeface. Among the attributes are face, set width, x height, ascenders and descenders, and serif and sans serif. The following sections discuss each of these attributes.

## THE MANY FACES OF TYPE

The face of a letter is the part that prints. An easy way to illustrate the term face is to think of the way in which stencils work. A sheet of stencils contains the *outline* of the letters of the alphabet. You place the stencil on paper, and then use a pencil to color in the *face* of the letter.

Faces are the thick and thin strokes that make up the letter. Chapter 1 discussed the six basic categories of typefaces. Type *style* is the way that the type face can be modified to add emphasis. Typically the style can be roman (normal), italic, bold, and bold italic.

Today's computers have added other options to type style. Not only can you specify roman, italic, bold, and bold italic, you also can designate the word to be underlined or to appear in strikethrough. Some programs may even give you the option of placing words in small caps or shadow text. Figure 2.8 illustrates typical type styles.

**Figure 2.8:** Examples of type styles of Times Roman.

Normal

**Bold**

*Italic*

***Bold italic***

Underlined

~~Strikethrough~~

Small caps

Shadow

## PROPORTIONAL TYPE VERSUS MONOSPACED TYPE

You can use two basic kinds of fonts: proportional and monospace. If you have no experience with fonts, you probably have used the Courier typeface on your computer. Courier is a monospace style of type. *Monospace* fonts give the same horizontal width to each character. A wide character, such as a "W" is given the same width as a narrow character, such as an "I."

Before computer fonts were developed, you had a choice of two monospace styles of type on your typewriter: pica and elite. These styles had more to do with the pitch. *Pitch* is the horizontal measurement of the characters. Pica pitch equals 10 characters per inch. Elite pitch equals 12 characters per inch. Don't confuse the term pica with the typesetting measurement system. The typesetting pica equals 1/6 of an inch.

*Proportional* faces, however, are designed to give each letter a width in proportion to the letter's shape. The words on this page are proportional. Monospace fonts take up more room than proportional fonts and look less professional (see fig. 2.9).

Typestyle

Typestyle

**Figure 2.9:** Monospace and proportional fonts.

**The Coach Says...**

Don't think that you should never use monospace fonts. If you want your reader to think that the document was typed on a typewriter, a monospace font is appropriate.

Monospace fonts also are used in computer documentation. These fonts most closely resemble the letters that you see on your computer screen.

Proportional typefaces also vary in their set width. Set width is the relative wideness or narrowness of the characters. Within the same typeface, you can choose a font with a relatively narrow width, or condensed face, or with a wide width, or extended face (see fig. 2.10). Fonts that have a narrow set width look as if a person has taken the top of the letter and stretched it to make it look taller and more narrow.

**Figure 2.10:** Set widths of typefaces vary.

Set widths vary.
Set widths vary.

**The Coach Says...**

Not all fonts have expanded or condensed versions. Many application programs enable you to alter the set width of characters, but there is a penalty in that some strokes of the characters become distorted.

## THE UPS AND DOWNS

All printed words rest on a *baseline*, which is an imaginary line on which all characters rest. The bottom of each character rests on this imaginary line. Lowercase letters that sit on this line may have some parts that rise above this line, and other parts that fall below this line.

The parts of the letter that rise above the body of the letter are called *ascenders*. Ascending letters are b, d, f, h, k, l, and t. The parts of the letter that fall below the baseline are called *descenders*. Descending letters include g, j, p, q, and y.

The *x-height* is the body height of lowercase letters. Some lowercase characters exist entirely within the x height. These characters include a, c, e, m, n, o, r, s, u, v, w, x, and z. These letters do not have ascenders or descenders.

Although x-height is not a unit of measurement, it does convey a visual impression of size. Some typefaces have a small x-height and others have a larger x-height. Figure 2.11 illustrates the word "exit" in four different typestyles. Although each word appears in the same size, some of the typefaces appear larger than others. Notice the lowercase letters. Some of the type styles have a larger x-height than others.

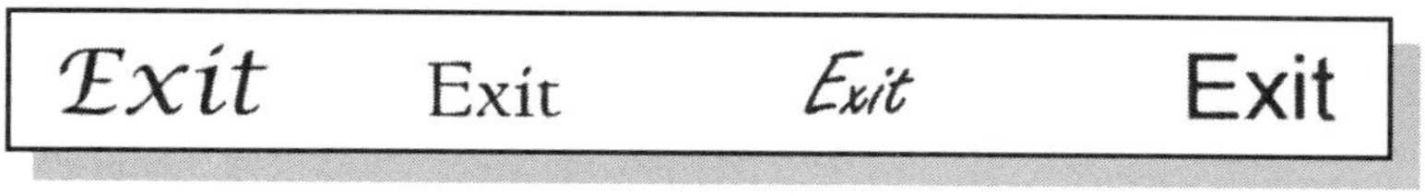

Exit Exit Exit Exit

**Figure 2.11:**
Type styles vary.

## SERIFS OR SANS SERIF

Look at the words on this page. Notice that at the ends of the lines that make up the letters are short stems. These stems are called *serifs*. The serifs tend to guide a reader's eyes across the page; therefore, choosing a serif style increases the readability of long passages of text. The word serif comes from the word scribe and means stroke or line.

Some fonts do not contain these strokes and are called sans serif. Sans serif faces contain no stems. *Sans* means without. Sans serif faces are most often used in headlines or in short documents. Figure 2.12 illustrates the difference between serif and sans serif.

**Figure 2.12:** Examples of a serif and sans serif font.

This is a serif face.
This is a sans serif face.

Some readers find it difficult to read long passages of sans serif type. You can determine whether this statement is true by performing the experiment in Practice Session 2.1.

**Practice Session 2.1**

**Judging Readability**

The word typography is shown in a serif and a sans serif face. Take a piece of plain white paper and place it over both words, aligning the top of the paper with the guides on either side of the words.

typography typography

Which typeface do you find easier to read? Most people find that the serif face makes for more comfortable reading.

## HOW BIG IS IT?

Now that you are more familiar with the anatomy of the individual characters, you can begin to look at the overall picture of typography. If you read any magazine or newspaper, you quickly see that

the fonts on the page vary in size. The headlines appear larger than the body text. Typography has its own system of measurement. This measurement system is much like using a typical inch ruler.

Measuring the size of fonts and the width of the printed line involves a measurement system that was developed in 1886. This system is based on picas and points. Approximately six picas equal one inch. There are 12 points in one pica, and 72 points in one inch (see fig. 2.13). You measure line length in picas and points; you measure the space between lines in points.

## The Coach Says...

Traditionally, 72.254 points have equalled an inch (0.128"), but most desktop publishing programs have standardized on 72 points per inch. Actually, the IBM Selectric Composer adopted this measurement back in the sixties.

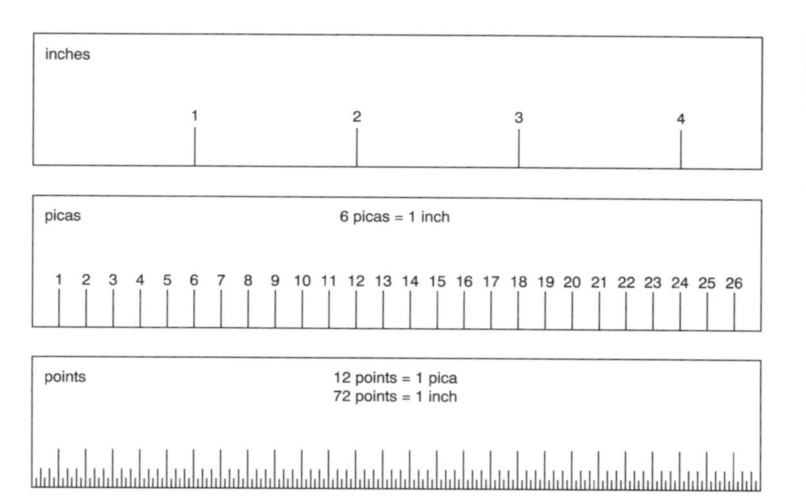


**Figure 2.13:**

The measuring system of typography.

## MEASURING POINT SIZE

Until the adoption of this measurement standard, typographers had no way to compare typefaces of different sizes. Each size of type had an individual name. Rather than specifying a type by size—type was specified by name. If you wanted to use a specific font—Gutenberg, for example—you could only use it as one size.

After the picas and points standard was adopted, a specific typeface could be developed in many sizes. All typefaces of a specific size, such as 72 point, were cast onto a one-inch metal block (which is 72 point). The blocks had to be of equal size so that the letters would line up properly. Because a letter does not necessarily fill the entire block, the actual image of the printed letter does not measure the same as the block (see fig. 2.14). A lowercase "a", for example, does not have an ascender or descender. Therefore, this letter does not fill the entire block.

**Figure 2.14:**

A block of type.

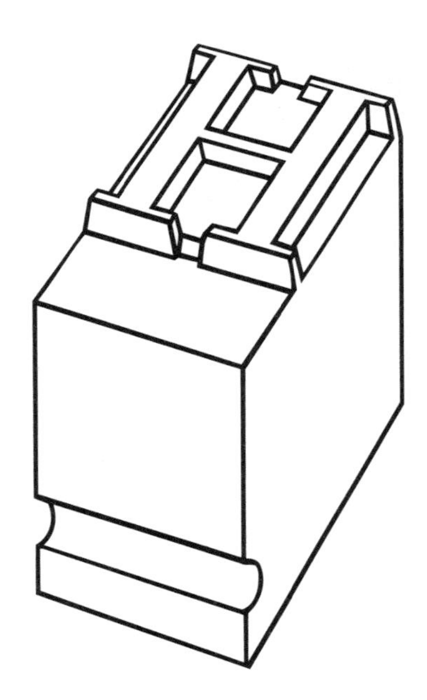

As you learned earlier in this chapter, the x-height of a typeface has a large impact on the appearance of the face. The larger the x-height, the larger the face appears. Figure 2.15 shows examples of three faces all in 30-point type. Notice that these letters have different heights. Caslon type seems larger than Technical and Howard because it has a much larger x-height.

Technical **Caslon** **Howard**

**Figure 2.15:** Three examples of 30-point type.

Measuring type with a pica ruler, therefore, is difficult. Type no longer is cast onto metal. You do not know the upper and lower baseline positions. If you measure from the top of an ascender to the bottom of a descender, therefore, you will only get an approximation of the point size. Fortunately, the computer takes care of this difficult job.

Today's word processing and desktop publishing programs enable you to use a wide assortment of type sizes. In the old days, a printer had to cast the typeface onto the metal blocks and print a sample copy to see what the type would look like. Now, however, you can try out typefaces and sizes just as you try on shoes. If you try one combination and don't like the result, simply try another combination. It's as simple as that.

### The Coach Says...

As you acquire more and more fonts, you may forget the specific fonts that you have, or you may forget what they look like in different point sizes. You can create a catalog of your fonts so that you have a

*continues*

*continued*

handy reference. This catalog of type sizes and fonts is called a *type specimen* book. Simply type a short paragraph of text, copy it several times, and then apply a different point size to each paragraph. Figure 2.16 shows a type specimen generated by using the TypeBook freeware program by Jim Lewis.

**Figure 2.16:**

A page from a type specimen book.

**Palatino-Roman**

6 pt. ABCDEFGHIJKLMNOPQRSTUVWXYZabcdefghijklmnopqrstuvwxyz0123456789!?,"¢$&%{}*
7 pt. ABCDEFGHIJKLMNOPQRSTUVWXYZabcdefghijklmnopqrstuvwxyz0123456789!?,"¢$&%{}*
8 pt. ABCDEFGHIJKLMNOPQRSTUVWXYZabcdefghijklmnopqrstuvwxyz0123456789!?,"¢$&%{}*
9 pt. ABCDEFGHIJKLMNOPQRSTUVWXYZabcdefghijklmnopqrstuvwxyz0123456789!?,"¢$&%{}*
10 pt. ABCDEFGHIJKLMNOPQRSTUVWXYZabcdefghijklmnopqrstuvwxyz0123456789!?,"¢$&%{}*
12 pt. ABCDEFGHIJKLMNOPQRSTUVWXYZabcdefghijklmnopqrstuvwxyz0123456789!?,"¢
14 pt. ABCDEFGHIJKLMNOPQRSTUVWXYZabcdefghijklmnopqrstuvwxyz012
18 pt. ABCDEFGHIJKLMNOPQRSTUVWXYZabcdefghijklmno

48 pt. ABCDEFGHIJKLMN
OPQRSTUVWXYZab
cdefghijklmnopqrstu
vwxyz0123456789!?,"
¢$&%{}*

9/12
This page was generated by "theType Book". A freeware power tool from Ji m Lewis. Customized versions are ava ilable in which the page footer is confi gured to your specifications. A typical layout would include a logo and comp any address information. Artwork is li mited to a height of 32 pts. The combin ed width of both copy and artwork is l imited to the printed area. The fee for c ustomization is $45(US). To order cont act Jim Lewis at the address shown bel ow.

10/12
This page was generated by "theT ypeBook". A freeware power tool f rom Jim Lewis. Customized versio ns are available in which the page footer is configured to your specifi cations. A typical layout would inc lude a logo and company address i nformation. Artwork is limited to a height of 32 pts. The combined w idth of both copy and artwork is li mited to the printed area. The fee f or customization is $45(US). To or der contact Jim Lewis at the addres

11/12
This page was generated by "th eTypeBook". A freeware power tool from Jim Lewis. Customiz ed versions are available in whi ch the page footer is configured to your specifications. A typica l layout would include a logo a nd company address informati on. Artwork is limited to a heig ht of 32 pts. The combined widt h of both copy and artwork is li mited to the printed area. The f ee for customization is $45(US).

| Point size: | 6 | 7 | 8 | 9 | 10 | 11 | 12 | 13 | 18 | 20 | 24 | 36 |
|---|---|---|---|---|---|---|---|---|---|---|---|---|
| Chars/Pica: | 4.45 | 3.82 | 3.34 | 2.97 | 2.67 | 2.43 | 2.23 | 2.06 | 1.48 | 1.34 | 1.11 | 0.74 |

| Cap Height: | 0.25" | 0.5" | 0.75" | 1" | 1.25" | 1.5" | 1.75" | 2" | 2.25" | 2.5" | 2.75" | 3" |
|---|---|---|---|---|---|---|---|---|---|---|---|---|
| Approx. pt. size: | 24.56 | 49.12 | 73.68 | 98.24 | 122.8 | 147.36 | 171.92 | 196.48 | 221.04 | 245.6 | 270.16 | 294.72 |

**Customized/Registered versions are available from**
Golden State Graphics • 2137 Candis • Santa Ana, CA 92807
714/542-5518 • CIS 71650,2373 • AOL JimXLewis

## MEASURING LINE LENGTH

Computer technology did not evolve out of the field of typography. This technology was originally in the hands of programmers and business folk who wanted to record data. They were not concerned with the aesthetics of a document, only with recording the data. Word processing programs, therefore, were set up more like typewriters than typesetting machines.

When you use a typewriter, you set the margins for your document, setting boundaries for your printed words. The typewriter allows you to type from one boundary to another boundary. Setting margins requires that you consider how much space you want on either side of your printed text.

In typesetting, however, you set *line length*. You consider how long you want the printed line of text to be. Measuring the line length of printed type is easy. If you can use an inch ruler, you can use a pica ruler.

## PICAS AND POINTS

Although popular word processing programs still use inches as the system of measurement, most desktop publishing programs give you the option of using picas and points. Figure 2.17 shows Word for Windows' ruler, which operates by default on the inch system. You can change to other measurement systems. Figure 2.18 shows PageMaker's measurement options. You may find it wise to switch to this system of measurement.

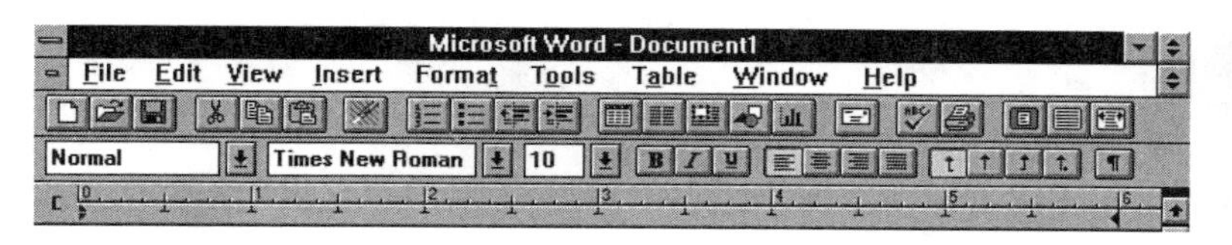


**Figure 2.17:** Word for Windows' ruler.

**Figure 2.18:**

PageMaker's measurement options.

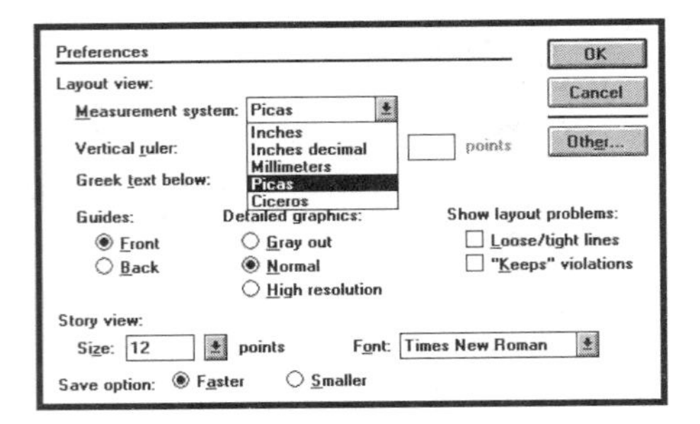


### The Coach Says...

If you plan on having your documents professionally printed, note that commercial quick printers operate on the picas and points system. If you learn the terminology now, you won't feel intimidated when you hear these terms bounced around. Measuring lines in picas is also much easier. If you have a mortal fear of math, you will appreciate the ease with which you can measure columns of text in picas. You don't need to figure complex fractions in your head.

## MEASURING THE SPACE BETWEEN LINES

The space that separates lines of text is called *leading*, which is pronounced "led-ding." A more common term for leading is line spacing. Leading is measured in points. You measure from the baseline of one line of type to the baseline of the following line of type (see fig. 2.19).

Leading is the space between lines. Leading is the space between lines. Leading is the space between lines. Leading is the space

**Figure 2.19:**

Leading is the space between lines.

### The Coach Says...

Line spacing is a more common term for leading in the computer arena. Desktop publishing programs, such as PageMaker, Ventura Publisher, and QuarkXPress, call it leading, however. If you are going to be doing desktop publishing, therefore, you should learn to use the term leading.

Leading is just as important to the look of your documents as choosing the font and the line length. If your lines of type are too close together, your document will look cramped and be difficult to read. If the lines are too far apart, your documents will look too airy and be difficult to read.

Think of leading in terms of how you used to use the spacing setting on your typewriter. You could single space, double space, or triple space your documents. Leading is very similar, but it gives you far more than three choices.

### The Coach Says...

For more information on the guidelines you should use when you set leading, see Chapter 8. The leading that you use can greatly enhance or detract from your document. Lines of type that are too close together are difficult to read.

You must consider each document individually before you adjust the leading. Think about what your goal is, who is going to read the printed document, and which typeface you want to use. As a rule of thumb, leading usually measures 20 percent more than the point size of the font that you are using. If you are using 10-point type, for example, an appropriate leading setting is 12 points.

> **The Coach Says...**
> The documentation and dialog boxes in many programs set automatic leading to 120 percent. This setting means the same thing as setting leading at 20 percent of the type size.

Typographers use a form of shorthand to specify point size, leading, and line length. You write the point size first, then a slash mark, and then the leading in points. To note the line length, you write an "x" and then write down the length in picas. The notation 10/12 x 20 actually means that the text is 10-point type on 12 points of leading set at 20 picas wide. Now when you hear your printer say "10 on 12," you will no longer wonder what he means.

## MEASURING WITH A PICA RULER

You use a pica ruler in much the same way as you use a typical inch ruler. The similarity ends there, however. A pica ruler looks very different than an inch ruler (see fig. 2.20). Pica rulers generally have picas marked along one edge of the ruler, and inches marked along the opposite side. The middle section of the ruler contains different measuring grids.

**Figure 2.20:**

A pica ruler.

### The Coach Says...

You can purchase a pica ruler from your local graphic arts supply store, or you can make your own. Simply photocopy figure 2.20 onto a transparency sheet. You will have your very own transparent pica ruler.

These grids reflect different leading settings. The section marked 10, for example, is marked off every 10 points. The column marked 12, is marked off every 12 points. These grids enable you to measure the leading of printed documents. You simply lay the pica ruler on top of a printed document. When the grid marks of the point settings align from baseline to baseline of the printed document, you can determine the leading. Figure 2.21 illustrates the way that you measure leading.

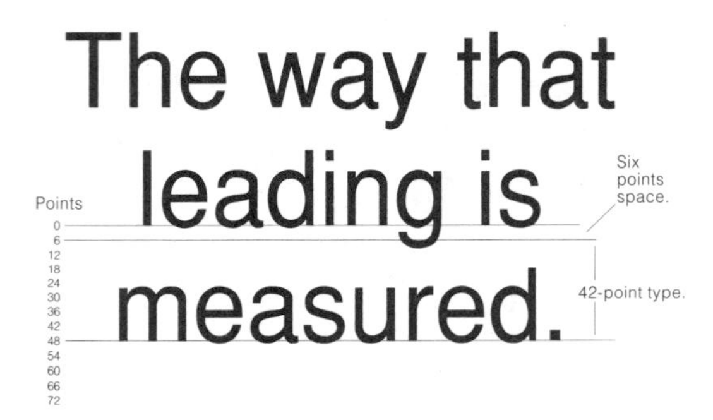


**Figure 2.21:** Measuring leading.

With a little practice, you too can use a pica ruler and measure line length and leading. Exercise 2.2 helps you understand how to measure the line length and leading of three passages of text. All type is 10 point.

## Practice Session 2.2
## Using a Pica Ruler

Use the pica ruler that you purchased or the ruler that you made, and measure the following passages of type. Fill in the blanks under each passage to record your calculations.

Use your pica ruler to measure this block of text. Record the point size of the type, the leading, and the line width in the spaces below. Use your pica ruler to measure this block of text. Record the point size of the type, the leading, and the line width in the spaces below. Use your pica ruler to measure this block of text. Record the point size of the type, the leading, and the line width in the spaces below.

Type size _______ points Leading _______ points
Line width _______ picas

Use your pica ruler to measure this block of text. Record the point size of the type, the leading, and the line width in the spaces below. Use your pica ruler to measure this block of text. Record the point size of the type, the leading, and the line width in the spaces below. Use your pica ruler to measure this block of text. Record the point size of the type, the leading, and the line width in the spaces below.

Type size _______ points Leading _______ points
Line width _______ picas

Use your pica ruler to measure this block of text. Record the point size of the type, the leading, and the line width in the spaces below. Use your pica ruler to measure this block of text. Record the point size of the type, the leading, and the line width in the spaces below. Use your pica ruler to measure this block of text. Record the point size of the type, the leading, and the line width in the spaces below.

Type size _______ points Leading _______ points
Line width _______ picas

How did you do? The first passage of text is printed in 10-point type set on 12 points of leading, with a line length of 13 picas. The second passage of text is 10-point type set on 14 points of leading, with a line length of 20 picas. The third sample is 10-point type set on 16 points of leading, with a line length of 22 picas.

This notation evolved from Gutenberg's day. The word *leading* is a holdover from the days when letters were individualy cast in lead. Lines of type were placed one after another in a form to construct a page. Thin strips of lead were inserted between the lines of type so that the text was easier to read. If no extra space was inserted between lines, the type was said to be set *solid*.

**The Coach Says...**

With computers, you now can set negative leading. You use negative leading to move lines even closer together than solid leading or to cause lines to overlap for special effects.

# INSTANT REPLAY

This chapter taught you the following skills:

INSTANT REPLAY

- ✔ A brief history of written language
- ✔ The development of the art of typography
- ✔ The anatomy of a letter
- ✔ The invention of movable type
- ✔ The picas and points measurement system

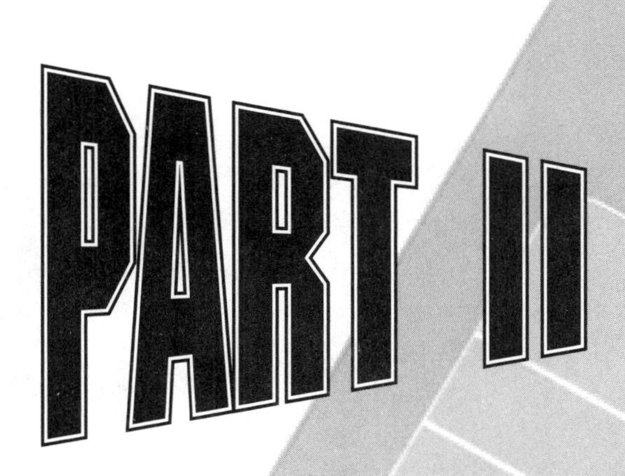

# PART II

# UNDERSTANDING COMPUTER FONT TECHNOLOGY

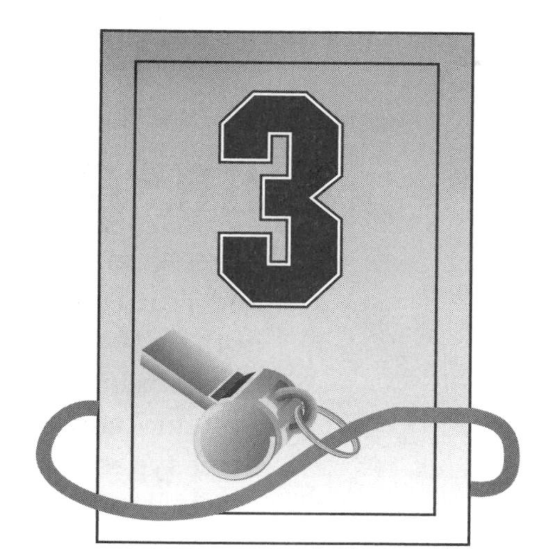


# UNDERSTANDING FONT TECHNOLOGY

So far, you have learned the Why, What, and Where of Fonts. In this chapter, you get into the nitty-gritty of How! This chapter teaches you the basics of the different technologies without overburdening you with technobabble. You get a quick discourse on what your font technology choices are along with their respective advantages and disadvantages.

Specifically, this chapter covers the following topics:

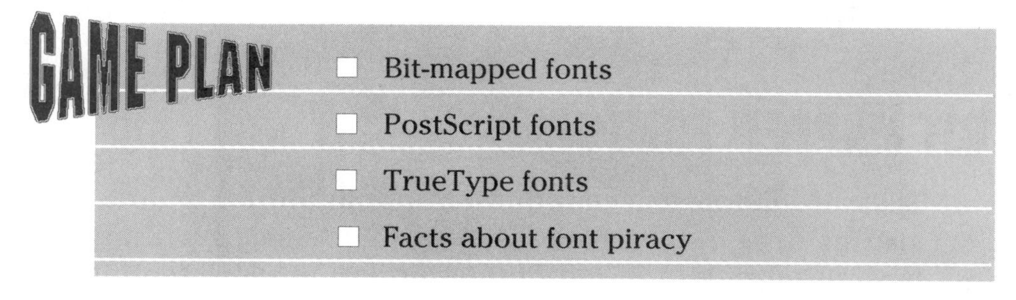


- Bit-mapped fonts
- PostScript fonts
- TrueType fonts
- Facts about font piracy

# A LITTLE FONT BACKGROUND

Three kinds of fonts are available: bit-mapped, PostScript Type 1, and TrueType. These are listed in the order of their appearance both on the market and in this chapter. Fortunately, these three can coexist on your system. One of the nice things about the way System 7 and Windows 3.1 handle fonts is that you can have all three types loaded and accessible at all times. You are not tied to one technology; rather, you can reap the benefits of each.

As you'll see, each font technology has pros and cons. You must determine what type of user you are—casual, business, or professional—and then decide which technology is best for you. As you read through the descriptions of each, think about your font needs. Ask yourself what is important to you: quality, price, selection, or ease-of-use.

While other font scalers are available—such as Bitstream's FaceLift for Windows and Zenographics' SuperPrint—this chapter focuses on PostScript and TrueType, by far the two most popular scalable font technologies.

This chapter gives you an overview of all three kinds of type. You'll find more in-depth information in the following chapters. Chapter 4 covers TrueType, and Chapter 5 delves into PostScript.

# BIT-MAPPED FONTS

The most basic of font choices, bit-map fonts are now archaic. Bit-map fonts can be thought of as a throwback to the days of metal type. Like their lead ancestors, bit-map fonts require a separate font for each specific point size. This hogs disk space unnecessarily, but can work for some users in a limited situation.

Why do bit-map fonts require a separate font for each point size? It's in the way that the font is built. Each character is defined on a pixel-by-pixel (screen dot-by-screen dot) basis. Take a look at figure 3.1; it shows how the letter A is defined in bit-map terms.

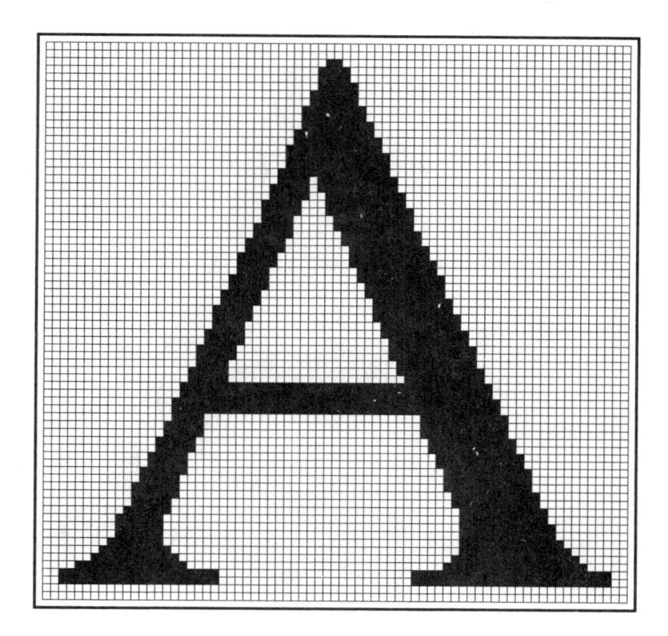

**Figure 3.1:**

Bit-mapped fonts are defined dot-by-dot.

Notice how the character is defined by the underlying grid. Now, compare figure 3.1 with figure 3.2, where the same bit-map letter A is shown without the grid. The jagged edges (called "jaggies") are way too obvious. That is why you don't want to use a bit-map font in a size other than its original.

Shrinking or enlarging fonts is known as *scaling*. When using bit-map fonts, always try to avoid scaling them; if you need a 12-point bit-map font, use a 12-point font. If you scale a smaller font up to size, you end up with jaggies. Figure 3.2 shows an enlarged version of what you would get if you scaled up a bit-map font. While this

illustration is a bit (no pun intended) exaggerated, you get the idea. Jaggies show the stair-stepped outlines that are telltale of scaled-up bit-map fonts.

**Figure 3.2:**

Bad jaggies!

Bit-mapped fonts come from many sources. It's likely that you have a few on your system, even if you don't realize that you do. Bit-mapped fonts are provided with a large number of printers and programs, and come with both Macintosh System software and Microsoft Windows. On the Mac, you can find a bunch of bit-map fonts that are provided courtesy of Apple. You can easily identify many of these by their names, which pay homage to the great cities of the world: Cairo, Chicago, Los Angeles, Monaco, New York, and San Francisco. Microsoft Windows bit-mapped fonts, on the other hand, can be identified by their FON filename extensions. While Microsoft was not as generous in providing a variety of bit-mapped fonts as Apple, enough fonts (such as the ever-present Courier) are available to handle the basics.

## ADVANTAGES OF BIT-MAPPED FONTS

Bit-mapped fonts provide two primary advantages: cost and speed. If you are a typographic unsophisticate, you can get away with using the fonts that are supplied with your Mac or Windows system. If all you do is straight word processing and you can live with just a few fonts in a number of predefined sizes, then you really don't have to go past bit-mapped fonts. In these cases, you'll get what you need, and you'll save a few bucks.

Screen display speed is a consideration for fast typists who use slow systems. Font-scaling technologies such as Adobe Type Manager and TrueType need to create bit-mapped versions of vector fonts for screen display; this process can become a drag on system resources. On slow systems (like old Macs and underpowered Windows PCs), you may actually wait for characters to display. This results in a clunky feel while you type. The more typefaces and point sizes you use in a document, the more overhead you create. This lag time can be unacceptable, but the options are obvious: either upgrade your machine or drop back to bit-map fonts.

If you are saddled with one of those aforementioned slow systems, you don't have to settle for just the fonts that came with your Mac or Windows. In addition to the standard issue fonts, there are oodles of public domain bit-mapped fonts available for both platforms. If you are a Windows user and have commercial PostScript fonts you want to use but your system resources won't allow, you can convert PostScript fonts to bit-map format. The Adobe Font Foundry utility, which comes bundled with Adobe, Agfa, and Monotype PostScript font packages, handles the conversion from PostScript Type 1 fonts to bit-map fonts. Font Foundry is a great way to get more typographic mileage out of an aging computer.

## DISADVANTAGES OF BIT-MAPPED FONTS

Bit-mapped fonts are yesterday's technology. Unlike scalable type (PostScript or TrueType), bit-mapped fonts need to have a font installed for every point size you use. As mentioned earlier, this quickly fills up hard disk space. The more fonts and sizes you use, the more disk space you lose. While you can scale a bit-map font up or down, it won't look right. Scaling a bit-map font up in size leads to jaggy type, which is a bane of desktop typography.

> **The Coach Says...**
> Third-party bit-mapped fonts contain differing numbers of point sizes. These sizes are determined by the manufacturer.

If you use bit-mapped fonts, you are tied to the font's resolution. The term *dots-per-inch (DPI)* is used to describe the resolution of a particular font or printer. Try to run a 300 dpi font on a 600 dpi printer, and you only get a 300 dpi font. As you may have guessed, more jaggy type! In short, the downside of bit-mapped fonts is their poor quality and inefficient use of disk space.

# POSTSCRIPT FONTS

As opposed to bit-mapped fonts, PostScript fonts are defined by mathematically-derived outlines. *PostScript fonts* use Bézier curves to create the character outlines. *Bézier curves* use mathematical formulas to describe a curve drawn between two anchor points. Each curve has a set of *tangent* (control) *points* to delineate the curves path. This methodology was developed in the early 1970s

by French mathematician Pierre Bézier. If this technology sounds like a bunch of gobbledygook from a nightmare math class, you need not worry about the particulars. You just deal with type as type, as usual.

The use of Bézier curves results in a totally scalable typeface that prints with smooth edges, no matter the point size. In other words: No jaggies! PostScript fonts are vector-based; that is, defined on a totally scalable, mathematical basis, as opposed to a matrix-based bit-map basis. Figure 3.3 displays the letter A on a grid, but this time it's a vector-based PostScript outline font.

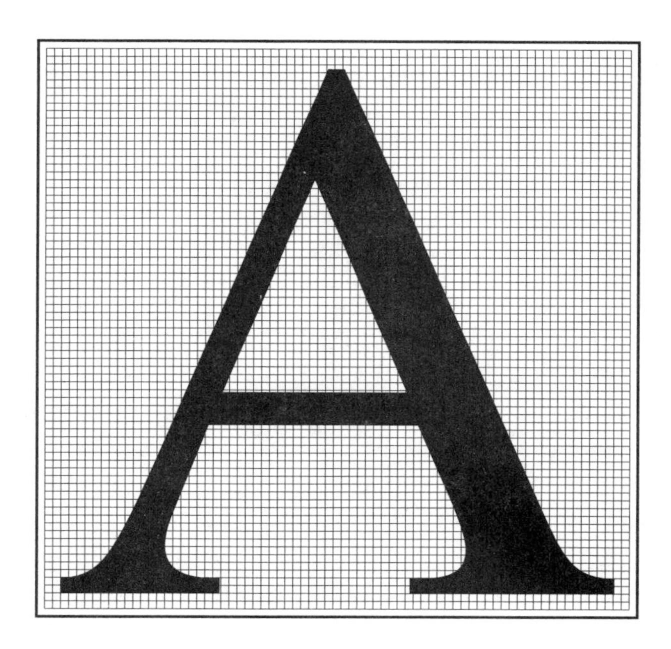

**Figure 3.3:** PostScript outlines are vector-based.

PostScript was first unveiled by Adobe Systems in 1985. It offered an exciting and innovative approach to the process of producing the printed page. What Adobe PostScript promised (and delivered) was printer independence, using scalable fonts and a common

*page description language*. In the early days, PostScript was unsuccessfully challenged by Xerox's Impress. Adobe won out, and the desktop publishing revolution was ignited.

Adobe's achievement of printer independence through a common page description language was monumental. Before PostScript, the world of fonts and printers was extremely proprietary. Back then, fonts from one foundry could not be used on a competitor's output device (whether it be a typesetter or printer). For example, you could not buy a font from Autologic and use it on a Linotype typesetter. An Autologic font could be used only on an Autologic typesetter, and a Linotype font could be used only on a Linotype typesetter. Each manufacturer had their own page description language. In this case, Autologic had its ICL format, while Linotype had its CORA format; and neither were interchangeable.

PostScript leveled the playing field. The printer independence that is now commonplace means that today, you can use that Autologic font on a Linotype machine (and vice versa). The PostScript page description language enables you to move pages from machine to machine without worrying about compatibility issues. If the page is setup in PostScript form, it can run on any one of scores of output devices from $1,000 laser printers to $100,000 imagesetters with the same integrity. Thus, you can easily and cost-effectively proof (and fine-tune) your pages on your desktop laser before you send them out to a service bureau for high-resolution output.

As you've learned, in PostScript's early days it used bit-map fonts for screen display (known as *screen fonts*) and the outline fonts only for printed output (known as *printer fonts*). With the release of Adobe Type Manager (ATM) in 1990, Adobe provided the capability to use the same outline font for both screen and printer. If you take another look at the letter A that you examined in the previous bit map section, you can see how PostScript eliminates jaggies from the character (see fig. 3.4). It does this by describing the character in vector, rather than bit-map terms.

**Figure 3.4:**

No jaggies with PostScript!

The next section takes a look at some of the advantages and disadvantages of PostScript fonts.

## ADVANTAGES OF POSTSCRIPT

PostScript fonts are the standard of the printing and publishing industries. Unlike TrueType fonts (which are discussed later), PostScript fonts are widely accepted at professional print shops and service bureaus around the world. PostScript imagesetting is a de facto standard.

PostScript Type 1 fonts are your choice for professional quality. They represent the best work of classic type foundries— Agfa, Berthold, Linotype, and Monotype to name a few—as well as

exciting new foundries like Émigré, The FontBureau, and Treacyfaces. Figure 3.5 displays the very popular Adobe Garamond family, one of the "Adobe Originals." In addition to Adobe Garamond, Adobe offers many other Garamonds: Berthold Garamond, Garamond 3, ITC Garamond, Simoncini Garamond, and Stempel Garamond.

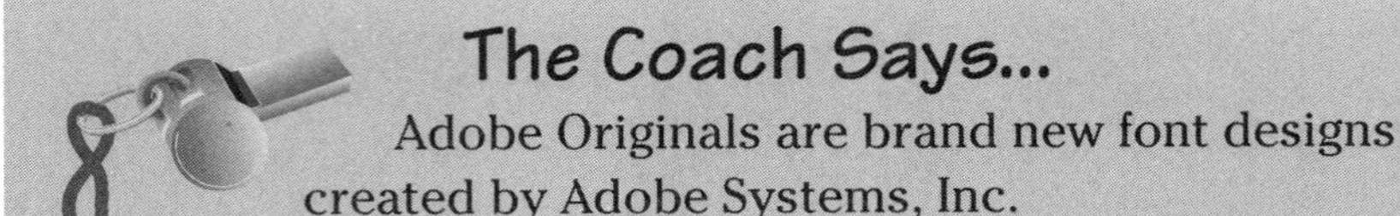

The Coach Says...
Adobe Originals are brand new font designs created by Adobe Systems, Inc.


**Figure 3.5:** The Adobe Garamond font family.

Adobe Garamond

*Adobe Garamond Italic*

**Adobe Garamond Semibold**

***Adobe Garamond Semibold Italic***

**Adobe Garamond Bold**

***Adobe Garamond Bold Italic***

By virture of variety alone, PostScript fonts are a clear advantage. A staggering volume of PostScript Type 1 typefaces currently are available—by some accounts 20,000 or more—from a wide variety of vendors. As of early 1993, Adobe's PostScript typeface library numbered over 1,200 fonts. Bitstream offers over 1,000 PostScript typeface selections, and The Font Company's Type Collection includes more than 1,600 fonts. (See Chapter 6 for more information on purchasing fonts.)

PostScript promises the exchange of files, not just between Windows PCs or PCs and Macs, but also between OS/2 (where PostScript is integral), Sun, NeXT, Silicon Graphics, and other platforms. As opposed to Microsoft's platform domination philosophy, Adobe's goal is platform and printer independence. With Adobe Type Manager, you do not need to have a PostScript printer. ATM enables you to print the PostScript fonts on any supported Windows output device.

## DISADVANTAGES OF POSTSCRIPT

The major disadvantage of PostScript fonts is monetary. To effectively use PostScript fonts, you must have Adobe Type Manager (usually referred to as ATM) loaded on your system. You do not need to have a PostScript printer, although if you do a good deal of graphics work, you'll want one (and you can expect to pay a few hundred dollars more for it). PostScript fonts, like TrueType fonts, may print more slowly than bit-map fonts.

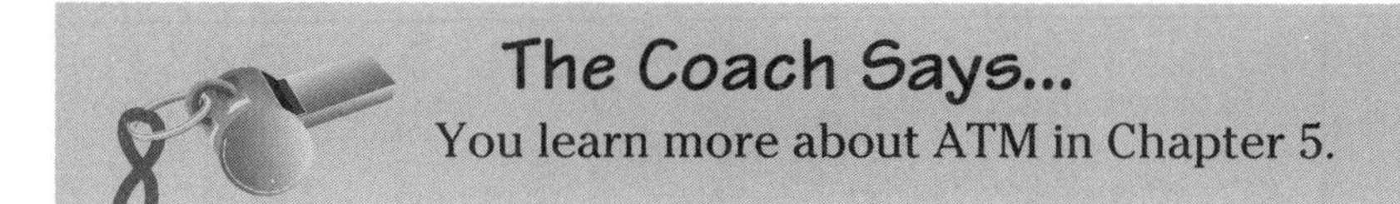

The Coach Says...
You learn more about ATM in Chapter 5.


The cost of ATM can be a consideration. While ATM is sold separately, it has been bundled with a wide variety of programs, such as Adobe Illustrator for Windows, Aldus PageMaker, Lotus 1-2-3 for Windows, and Micrographix Designer. ATM is upgraded regularly. Most of these upgrades are relatively inexpensive and commonly offered with attractive font bundles. Consequently, you pay for the upgrade but get a bunch of new, high-quality fonts at the same time. Are you paying for the upgrade or the fonts? It doesn't matter. ATM upgrades are a nice way to affordably augment your font library.

Most high-quality PostScript fonts are, as you might expect, relatively expensive. Plenty of low-cost (and lesser quality) fonts are available, and you can find bargains. Adobe frequently offers special pricing on certain font packages, as well as the aforementioned bundles. Figure 3.6 shows the 16 fonts in Adobe's New Faces Collection, available for $39 (direct) including the ATM 2.5 upgrade.

**Figure 3.6:** Adobe's New Faces Collection.

**Adobe's New Faces Collection**

| | |
|---|---|
| Ellington® Regular | Goudy Modern |
| *Ellington Italic* | *Goudy Modern Italic* |
| **Ellington Bold** | **Falstaff** |
| ***Ellington Bold Italic*** | **Forte** |
| Grotesque Regular | Klang® |
| *Grotesque Italic* | ***Mercurius™ Bold Script*** |
| **Grotesque Bold** | *Monoline Script™* |
| **Grotesque Bold Extended** | Monotype Old Style™ Bold Outline |

PostScript is no speed demon, however. You can expect to wait a while for some PS pages to print; this might or might not be a consideration depending on the type of work that you do. Adding more printer memory helps, as (obviously) does upgrading your printer to a faster and more powerful machine. In general, PostScript users have accepted the fact that print times can be lengthy, although they lust after faster output devices. Today's leading edge printers are considerably faster than the early designs.

# TRUETYPE FONTS

TrueType font technology is the result of the unlikely collaboration of (otherwise rivals) Apple and Microsoft. Apple began work on TrueType as a reaction to the success of Adobe PostScript. Microsoft soon bought into the technology. TrueType first appeared in Apple's System 7 in mid-1991. It was languishing until the Spring 1992 introduction of Microsoft Windows 3.1. With Microsoft's marketing muscle behind it, TrueType brought fonts to the business computing masses.

**The Coach Says...**

Windows 3.1 shipped with a handful of TrueType fonts. You will find the sans serif Arial font family (roman, italic, bold, and bold italic), along with the serif Times New Roman family (also in roman, italic, bold, and bold italic). The standard Windows 3.1 font load also includes the typewriter look-alike Courier New font family (once again in roman, italic, bold, and bold italic), Symbol (Greek) font, and the funky WingDings dingbat font.

Like PostScript, TrueType is a scalable font technology. This means that unlike bit-map fonts, you need only one font to create any point size type you want.

## ADVANTAGES OF TRUETYPE

A major advantage of TrueType fonts is that they are integral to Apple's System 7 and Microsoft Windows 3.1. You don't have to own (and constantly upgrade) another font scaler, such as ATM. If you have Windows 3.1, you can use TrueType. Also, DOS users are slowly gaining the capability to use TrueType.

TrueType fonts generally are less expensive than PostScript fonts due to differences in marketing philosophy. Microsoft's aggressive pricing strategy, however, has forced Type 1 prices downward. Microsoft's first TrueType package includes a good number of faces at an attractive price.

Another advantage of TrueType fonts is that they are not printer dependent. You can print TrueType fonts to any Windows-supported output device. In addition, they easily can be translated into PostScript fonts with FontMonger or Fontographer. (For more information on converting fonts, see Chapter 9.)

While TrueType fonts print to any Windows-supported graphics printer, they are somewhat printer dependent when it comes to using a PostScript printer. TrueType fonts are rasterized for the printer at the resolution specified in the Advanced Options of PostScript printer setup in Windows. If your driver is set to print to an imagesetter at 1200 dpi, but you print a proof to a 300 dpi laser printer, you end up with bold/chunky fonts. Conversely, if your printer driver is set to 300 dpi and you prepare a file to send to a service bureau for hi-res output, the fonts appear jagged (at 300 dpi resolution). When you use TrueType on a PostScript device, you must be sure to select the proper printer and set the proper resolution.

## DISADVANTAGES OF TRUETYPE

Relatively few professional-quality fonts are available in TrueType format, although this is likely to change with time. As of this writing, the majority of TrueType fonts—with the exception of a handful of collections—are knock-offs from bargain-basement font mills. The major foundries have been slow to create TrueType versions of their libraries. Most manufacturers have marketed small collections, however (see table 3.1). You can use both

PostScript and TrueType in a document, but print times might suffer.

**Table 3.1**
**A Handful of Quality TrueType Collections**

| *Foundry* | *Collection* | *Number of Fonts* | *List Price* |
|---|---|---|---|
| Agfa | Desktop Styles | 39 | $79 |
| Bitstream | TrueType Font Pack # 1 | 40 | $79 |
| | Type Essentials | 13 | $109 |
| Microsoft | TrueType Font Pack 2 for Windows | 44 | $70 |
| Monotype | ValuePacks | 57 | $89 |

You should remember that the output device—whether a desktop laser printer or a high-end imagesetter—cannot actually print fonts as outlines. Instead, these outlines must be *rasterized* (converted from vector-based outlines into a bit-mapped matrix pattern, at the resolution of the output device) to be printed because the printer's engine can lay down only dots, not lines or curves. This rasterization (or *ripping*, to those in the know) is done at the printer's Raster Image Processor (commonly referred to as its *RIP*). Most laser printers have internal RIPs, while most imagesetters have external RIPs.

To run a TrueType font on a PostScript RIP, a number of options are available. In the case of Microsoft Windows, these options are handled by the Windows PostScript printer driver. The procedure is by no means idiot-proof. You have to know what option to use, and how to set up the driver for optimum results. The three options are as follows:

★ **Substitute a printer-resident PostScript font.** This process is done via a font substitution table. Arial, for example, will print as Helvetica. While these typefaces might seem close in appearance, they are, in fact, different typefaces. Using font substitution means that you lose true WYSIWYG; in short, what's on your screen no longer matches what comes out of your printer.

★ **Convert TrueType fonts to PostScript Type 1 fonts.** If you select this option, the printer driver actually changes the TrueType font into a PostScript Type 1 font for the purpose of printing. Reports are that the results of this method are acceptable on high-resolution output devices.

★ **Convert TrueType fonts to PostScript Type 3 bit-map fonts.** This provides high-quality results, because the print driver builds each font for each specific point size. There is a downside, however; this method results in longer print times and bigger print files. In addition, the printer driver must be set up for the specific resolution of the output device. If you want to run the job at 2,540 dpi on a Linotype imagesetter, you must set up the printer driver with those exact specifications.

Because TrueType was developed by both Apple and Microsoft, other operating platforms have not rushed to embrace the technology. At present, the technology appears tied to these two operating environments.

Service bureaus have been hesitant to deal with files that contain TrueType fonts, as have color prepress houses. Some bureaus refuse to accept jobs with TrueType fonts. In addition, a widely accepted fact is that files containing TrueType fonts rasterize more slowly into high-end systems. No alternative, however, is available because no high-end TrueType RIPs exist (as of early 1993).

**The Coach Says...**

Service bureaus are unwilling to reinvest in new TrueType font libraries because of their present investment in PostScript font libraries, and are unlikely to do so until the market demands that they convert to this technology.

# FONT PIRACY

You might be a font *pirate* and don't even know it. Have you ever copied a typeface to give to a friend or co-worker? If so, you might have broken the law. Why? Take a look at the font license that comes with your commercial typeface. In most cases, when you buy a font you have not actually purchased the font itself. Rather, you have bought the rights to use the font in very defined terms. You don't really own the font. It might sound funny, but it's the law. This law is consistent with the way that most software is customarily distributed.

The typographic industry is very aware of the problems of bootleg fonts. An organization known as the Association Typographique International (ATypl) is sponsoring a Font Software Anti-Piracy Initiative. The following is the ATypI policy statement on the issue of font piracy:

> The use of a package of font software is governed by a license agreement. When font software is purchased, the rights the user has licensed do not include the right to make unauthorized copies of the type design or of the font software that embodies the design. If copies of the font are made to give away or resell, everyone involved in the creation of the font software, including the typeface designer, will be prevented from being properly rewarded for the hard work involved in

its creation. This could discourage the creation of new typefaces, hinder font software development, and reduce the ability of manufacturers to make new products available.

You have to police yourself on this issue. Bootleg fonts hurt everyone. If you're caught with pirated fonts, you or your company could be subject to legal action. The software industry loses millions of dollars a year from illegal copying. They are taking action by performing company audits. While this may conjure up images of 1920s speakeasy raids, it's a reality. Software companies are going to court, and consistently winning, with cases against bootleggers. Protect yourself, your company, and the typeface developers by not making or accepting illegal copies.

Taking your files to a service bureau (for high-resolution or color output) raises a question: how do you legally deal with the fonts? Quite honestly, it's chiefly the responsibility of the bureau. Better service bureaus own extensive PostScript font libraries. So when you bring in a file to print out, you shouldn't have to worry about bringing the fonts with you. Just make sure that they know which fonts you used in your file. Most bureaus have a standard form which they require you to fill out before your file is run. The font list is one of the most important items on that form. They don't want to waste anyone's time (or money) by running jobs that are missing fonts.

If you are running a file at a service bureau, and that bureau does not have the fonts you used in the file, you must supply the fonts with the job. However, these fonts are supplied to run that particular job, and no one else's job. The bureau should not download those fonts to their printer's hard disk, nor should they store them on a workstation. In addition, they should return your disk (containing those fonts) when they deliver your job. Legally, they do not have the license to hold a copy of a font that they do not own.

Perhaps the cleanest way around the issue is to provide the bureau with PostScript print files rather than application files. When you create a PostScript print file, the printer driver actually loads the printer font file into the PostScript print file. This way, you avoid having to hunt around your disk for the fonts, and the bureau saves time by not having to manually download them before printing the file from the application. Instead, the bureau simply sends the file down to the output device with a downloader utility. While creating a print file might seem like an extra step for you, it speeds the overall process and can save you a few dollars. Many bureaus offer a discount on jobs run from print files! Most applications have a way to prepare PostScript print files—check your documentation for details.

## INSTANT REPLAY

In this chapter, you learned about the three most common font technologies and the advantages and disadvantages of each.

The decision about which font technology to use largely depends on the work that you do. In most cases, you should be working with scalable fonts, such as PostScript or TrueType, as opposed to bit-map fonts. Whether you decide upon either of those technologies can be determined by how much you are willing to invest, and how much you are willing to compromise.

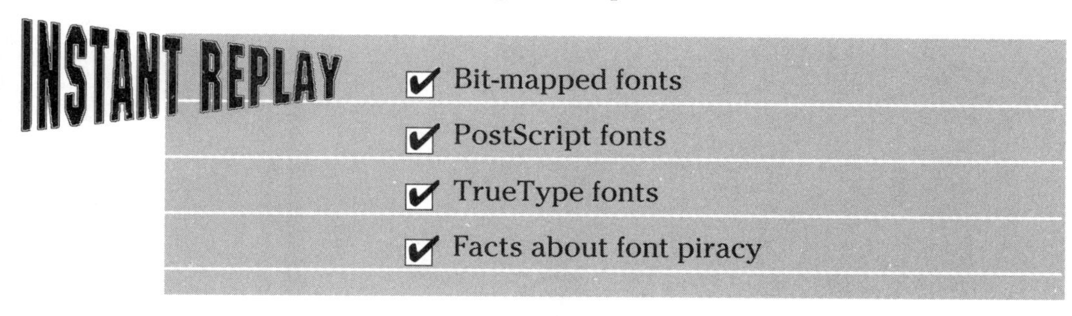

INSTANT REPLAY
Bit-mapped fonts
PostScript fonts
TrueType fonts
Facts about font piracy

# UNDERSTANDING TRUETYPE

If you are new to the world of fonts, your first excursion may be to test drive the TrueType fonts that come with Windows 3.1 or Macintosh System 7. Fans of TrueType applaud its availability and ease of use, and claim that it's the best invention since sliced bread. To separate the hype from the type, however, you need a basic understanding of this scalable font technology.

This chapter discusses:

## GAME PLAN

- The birth of TrueType
- What is TrueType?
- Hinting
- Installing TrueType on Windows and Macintosh
- Removing TrueType fonts from Windows and Macintosh

# ANOTHER SCALABLE FONT FORMAT?

You may be wondering why the computer industry developed yet another font format. Wasn't the technology complex enough without adding another format to worry about? Fonts took up a lot of storage space on your computer, and didn't always print out the way they looked on-screen. TrueType evolved partially because of concerns such as these and in response to Adobe's dominance in the PostScript Type 1 font arena.

## A COALITION IS BORN

TrueType is a font technology originally developed by Apple Computer. Apple introduced TrueType at a developer's conference in May 1989 as part of its new System 7 operating system. Apple's goal was to supply a font format that all Macintosh users could use and print. Microsoft, in the meantime, was working on its PostScript-compatible interpreter called TrueImage.

Apple and Microsoft shocked the computer community in September 1989 when the companies announced a joint venture to combine the TrueImage technology with TrueType technology. Apple would supply the font architecture, and Microsoft would supply the PostScript-compatible interpreter. Together, the companies could break the monopoly held by Adobe Systems.

## RESPONSE TO TRUETYPE

Eventually, Apple released TrueType fonts with System 7, and Microsoft shipped TrueType fonts with the release of Windows 3.1. A font explosion began.

Never before had a scalable font technology been available to so many people. TrueType had the added benefit of being easy to install and use. TrueType solved many of the compatibility problems and enabled users to print documents on a wide variety of printers.

Owners of PostScript-based printers were concerned at first. They wondered whether TrueType would take over as the dominant font format. Because Apple and Microsoft decided to make TrueType an open architecture, however, the technology actually opened up font competition. The companies realized that making TrueType available to type manufacturers would ensure that a wide array of fonts would be available to the public.

After the release of TrueType, Adobe in response to the competition, opened up its formerly proprietary Type 1 format to other type vendors. The price of Type 1 fonts actually fell.

### The Coach Says...

Most type vendors offer fonts in both TrueType formats and PostScript formats. Well-known type foundries such as Agfa/Compugraphic, Bitstream, Linotype, and URW now offer TrueType fonts that compare with PostScript.

Today, TrueType technology coexists with PostScript. You can use TrueType and Type 1 fonts not only on the same computer, but even in the same document. The existence of the two formats has caused a price war in not only fonts, but printers as well.

TrueType is here to stay. With the popularity of Windows 3.1, more and more computer users will be using TrueType fonts. Currently, more than 2,000 TrueType fonts are available.

# WHAT IS TRUETYPE?

TrueType is a *scalable* font technology. Each character in the character set is based on an outline, enabling you to change, or scale, the size of any character in the set. This single outline is responsible for what you see on-screen and what you see on the printed document. You need just one font file. Unlike bit-mapped fonts, TrueType fonts don't require a separate screen font and printer font file.

**The Coach Says...**

See Chapter 3 for a complete discussion and comparison of the various font formats.

## TRUETYPE SIMILAR TO POSTSCRIPT

Both TrueType and PostScript fonts use mathematical representations to reproduce the letters in a particular font set. Both font types use a font rasterizer to scale the letters and convert the outline into a bit-mapped letter. Simplified, this process means that the rasterizer reads the outline, draws the character in the specified size and style, and fills in the outline to produce the character.

**The Coach Says...**

Rasterize is the process that the font technology uses to convert a mathematical formula of the outline of a letter, and places the letter as a bit-mapped image on the screen or printed on paper.

PostScript fonts use a third-party rasterizer, such as Adobe Type Manager. TrueType fonts, however, have a built-in font rasterizer. Although both formats are scalable font technologies, the way that they go about scaling is different.

## HINTING

Computer screens are made up of a grid of dots called *pixels*. The higher the number of dots in a given area, the higher the quality or *resolution* of the image. Computer screens have a lower resolution than laser printers. The displayed image, therefore, contains a different dot pattern than the printed image. A typical computer monitor may display images at 96 dots per inch, but a typical laser printer prints images at 300 dots per inch. Professional typesetting machines may print images at 1,200 dots per inch. Figure 4.1 illustrates the difference that can occur between screen images and printer images.

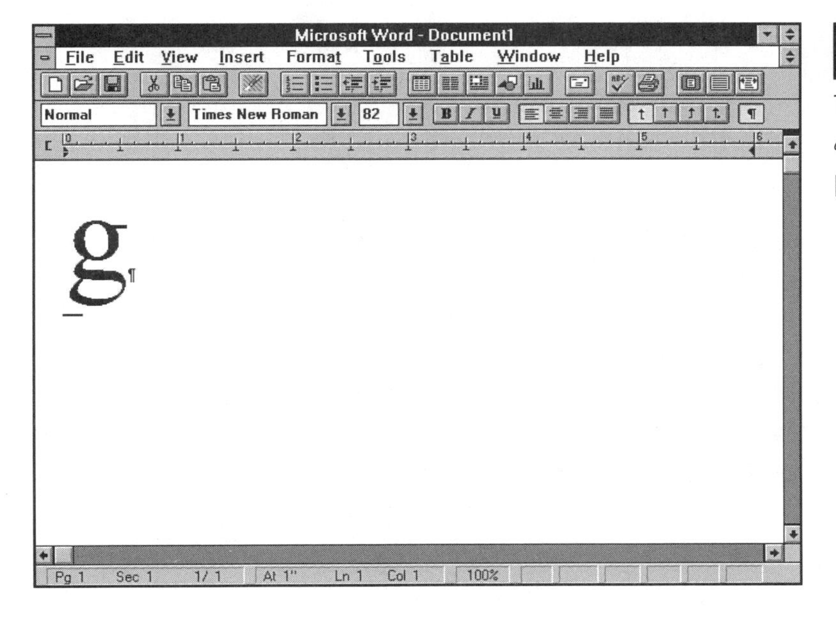


**Figure 4.1:** The screen display and printed display.

g

To compensate for this difference, scalable fonts use a technique called *hinting*. Hints are instructions from the font rasterizer that make adjustments in the dots that make up the character. The hints make up for the lower number of dots that are available on-screen.

Hinting instructions make sure that letters appear correctly at different sizes. Hinting also helps preserve the shape of a character at different resolutions. Figure 4.2 illustrates a letter before and after hinting. Note that the first stroke of the letter "n" is three pixels wide, but the second stroke is only one pixel wide. The width of the strokes is dependent on where the strokes fall in the grid pattern of pixels. Hinting makes adjustments to the character so that it appears more even (see fig. 4.2). Hinting reduces the jagged edges of the image.

**Figure 4.2:**

The effects of hinting.

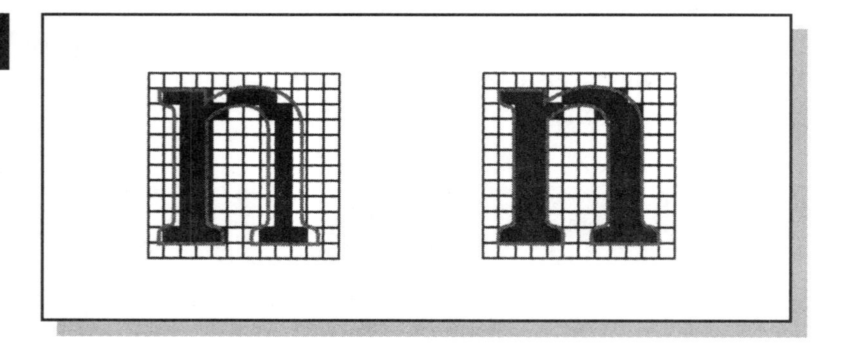

The smaller the point size you choose to use, the smaller the number of available pixels to display the character. Hinting, therefore, becomes quite important at smaller point sizes (below 8 point). Delicate features of letters, such as serifs, can become distorted because the strokes take up more pixels in proportion to the rest of the character. Hinting guarantees that all features of a letter are displayed in the correct size and proportion.

## DIFFERENCES IN TRUETYPE AND POSTSCRIPT HINTING

TrueType fonts differ from PostScript fonts in the way that they handle hinting. PostScript fonts use a *declarative hint*, which tells the rasterizer the amount that a character can be distorted. The rasterizer implements the hints.

With TrueType's built-in rasterizer, the character is displayed as closely as possible to the original shape. TrueType uses a process called grid-fitting. Grid-fitting enables TrueType to hint diagonal lines to reduce the jagged effect of stair-step lines. Storing the rasterizer in the font also enables different fonts to treat characters in different ways.

TrueType also uses a different type of curve to construct letters. PostScript uses a curve called a Bézier, or cubic curve, to construct letters. TrueType, however, uses quadratic curves. Quadratic curves require more control points than cubic curves and produce a smoother curve.

Both PostScript and TrueType claim to have the best hinting capabilities. As an end user, you probably don't care about which type of hinting is applied to the font. You just want a good-quality look. The quality ultimately resides in the expertise of the font developer, not in the font technology itself.

### The Coach Says...

Note that the location of the font rasterizer does have an effect on future quality of your fonts. Having a separate font rasterizer with PostScript fonts does enable you to upgrade all fonts at once by upgrading to a newer version of your rasterizer. With TrueType, you must upgrade each font.

# SEEING WHICH TRUETYPE FONTS ARE ON YOUR SYSTEM

Windows 3.1 and Macintosh System 7 include several TrueType fonts. You easily can see the installed fonts on both systems. With the Macintosh, you can open the System folder to see the installed TrueType fonts. With Windows 3.1, you can consult the Fonts Control Panel.

## THE WINDOWS FONTS CONTROL PANEL

The Fonts Control Panel is part of the Windows Control Panel that usually is part of the Main program group window. From the Fonts Control Panel, you can view and install TrueType fonts. The following steps show you how you can see which fonts are already installed on your system.

1. If the Main program group window is not open, double-click on the Main icon (or the group in which the Control Panel application resides).
2. Double-click on the Control Panel icon (see fig. 4.3).

   The Control Panel is displayed.

**Figure 4.3:** The Control Panel icon.

3. Double-click on the Fonts icon (see fig. 4.4).

   The Fonts dialog box is displayed.

**Figure 4.4:**

The Fonts icon.

## TYPES OF FONTS

The names of the fonts that are installed on your system appear in a box in the upper left of the dialog box. The fonts are listed alphabetically within this box. Figure 4.5 shows the Fonts dialog box and installed fonts.

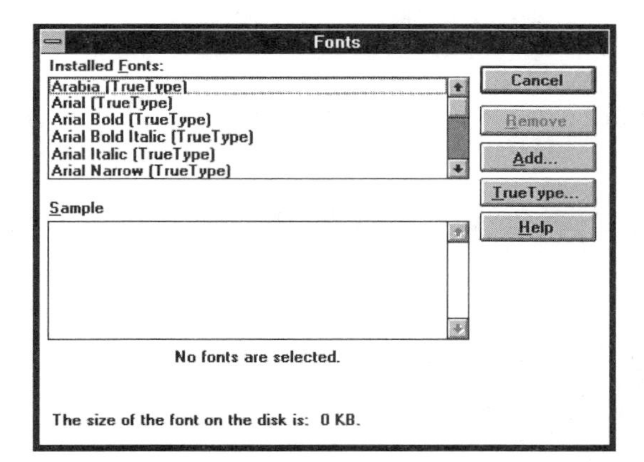


**Figure 4.5:**

The Fonts dialog box.

Notice that after the font name the type of font appears in parentheses. The font type can be screen, plotter, or TrueType. Screen fonts are those fonts that Windows uses to display the text that you see in dialog boxes and menus. Screen fonts are indicated by the type of graphics adapter card that you have. Plotter fonts are used with Computer Aided Design (CAD) software that sends output to plotters.

Windows 3.1 includes 14 separate TrueType fonts. These fonts are divided into three separate font families and two specialty fonts.

## LOOKING AT FONTS IN THE FONTS CONTROL PANEL

You can do more than just view the names of your installed fonts—you actually can see the way the font looks. The process is easy. Just move the pointer to the font that you want to view and click on the name. The font is displayed in the **S**ample box. Note that all TrueType fonts appear in the **S**ample box in 24 point size (see fig. 4.6). Beneath the **S**ample box, Windows displays the message `This is a scalable TrueType font that can be displayed on the screen and printed on your printer`. The file size of the TrueType font is also displayed.

**Figure 4.6:**

*Viewing a TrueType font.*

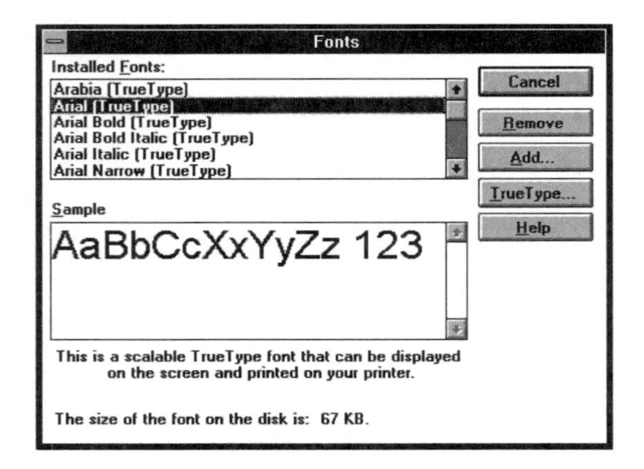


If you select a screen font, such as Symbol, the **S**ample box displays the font in all the listed point sizes (see fig. 4.7). Windows displays the message `This is a screen or plotter font` underneath the Sample box. Just as with TrueType fonts, Windows displays the font's file size beneath the Sample box.

## TRUETYPE FONTS ON THE MACINTOSH

The names of the fonts that are installed on your Macintosh system appear in your System folder. To see which fonts are currently installed on your system, simply double-click on the System file within the System folder. Make sure that you have selected to view the contents of the file by icon.

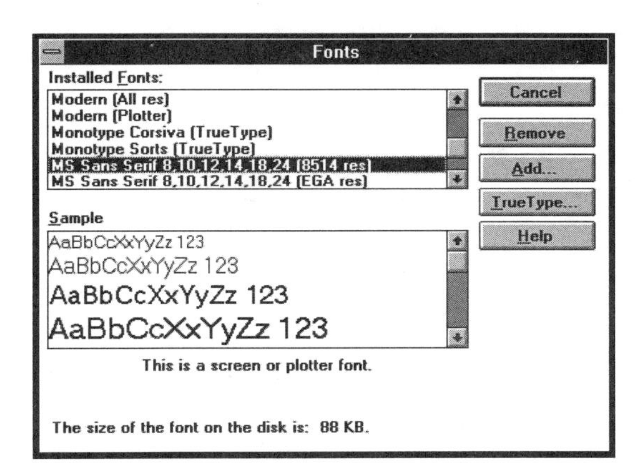


**Figure 4.7:** Displaying a screen font.

### The Coach Says...

The Macintosh enables you to see the contents of folders by name, size, kind, label, date, and two icon views. Changing the view is easy. From the View menu, select Icon. The contents of the folder are displayed as small pictures.

Because the Macintosh stores both fixed-size fonts and scalable TrueType fonts, you will see two types of icons (see fig. 4.8). The TrueType icon does not show a point size in the name. The fixed-size font shows the name of the font plus the actual size.

Figure 4.8:

TrueType and fixed-size icons.

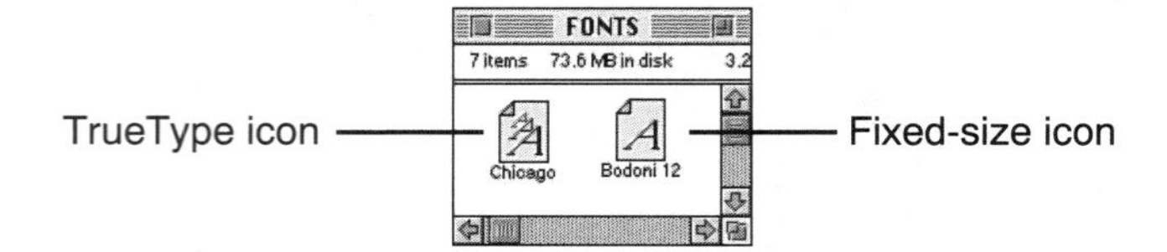


You will see just one TrueType icon for each font, but you may see many icons for fixed-size fonts. Figure 4.9 shows the Times Roman TrueType icon and the various icons for the fixed-size Helvetica font.

Figure 4.9:

TrueType requires only one icon.

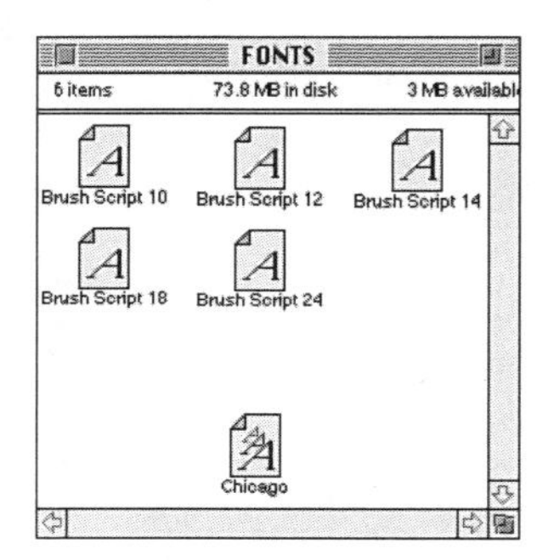


## SEEING THE FONTS

Seeing exactly what a font looks like is a simple procedure. Just double-click on the font that you want to see. A window is displayed and shows a sample sentence in the chosen font. If you double-click on a TrueType font, the window shows three sample type sizes. If you click on a fixed-size font, the window shows the size of the font that you chose. Figure 4.10 illustrates the way in which the Macintosh displays TrueType and fixed-size fonts.

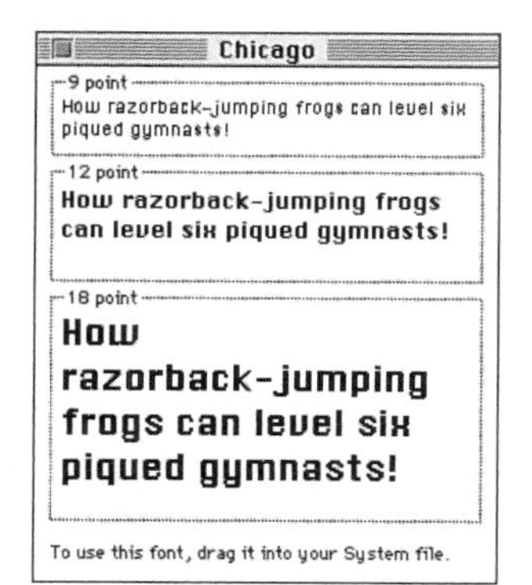


Fixed-size fonts

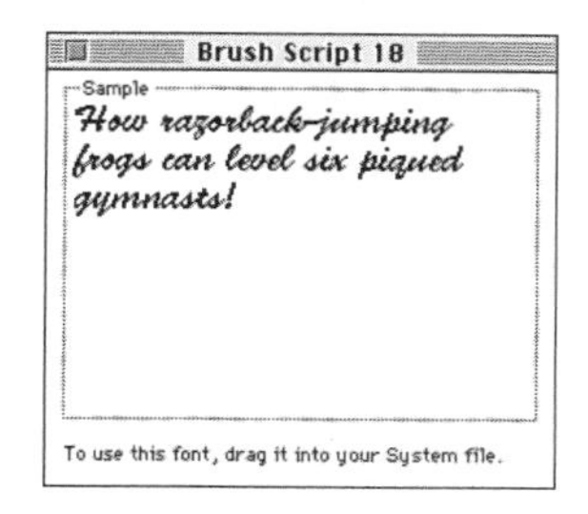


TrueType fonts

**Figure 4.10:** Displaying TrueType and fixed-size fonts.

**The Coach Says...**

The Macintosh can have both the TrueType version and a fixed-size version of the same font installed. Note, however, that if you choose a particular font, the Macintosh checks to see whether you have a fixed-size version of that font. If a fixed-size version is not available, it selects the TrueType version.

# INSTALLING TRUETYPE FONTS

Installing TrueType fonts is a much easier process than installing PostScript fonts. The process that you follow is similar for both Macintosh and Windows installation. After you install a TrueType font, you can use the font in any Windows or Macintosh application. You also can print TrueType fonts on any printer that supports TrueType.

## INSTALLING TRUETYPE FONTS IN WINDOWS

Some font packages include an installation program that enables you to install the programs on your computer without going

through the Fonts Control Panel. If you purchase the Font Pack for Windows from Microsoft, for example, the package comes with its own setup program.

Most of the time, however, you will be installing fonts from a floppy disk. In this case, you use the Fonts Control Panel. The Fonts Control Panel by default stores your fonts in the WINDOWS\SYSTEM directory.

> **The Coach Says...**
>
> Storing the fonts in the WINDOWS\SYSTEM directory is appropriate, if you do not have too many fonts. Remember, however, that each font takes up around 75K of storage space and memory. Storing large quantities of fonts may require a more advanced management system. Chapter 6 explains the procedure for storing and managing large numbers of fonts

Intalling fonts from a floppy disk is easy. Follow these steps:

1. Insert the disk that contains the TrueType font that you want to install into drive A or B.
2. In the Control Panel window, double-click on the Fonts icon.

   The Fonts dialog box is displayed.
3. Choose the Add button.

   The Add Fonts dialog box is displayed (see fig. 4.11).
4. Select the drive that contains the disk with the TrueType font that you want to install by clicking on the Drives down-arrow button.

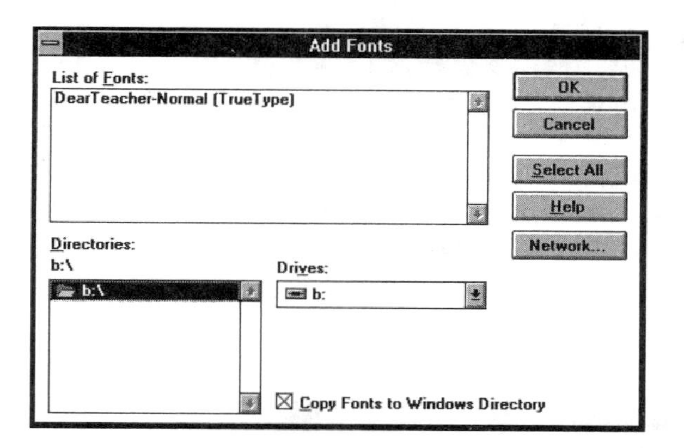


**Figure 4.11:** The Add Fonts dialog box.

You see the message `Retrieving font names`. As Windows reads the fonts, you see the percentage of the font names found. If the disk contains several TrueType fonts, this procedure may take some time. After all the fonts from the disk are found, the font names appear in the List of Fonts box in alphabetical order.

5. Select the fonts that you want to install. You can select all fonts from the disk by clicking on the Select All button (see fig. 4.12).

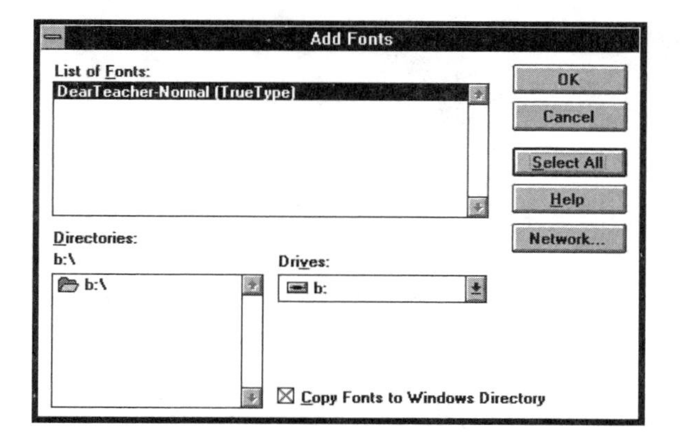


**Figure 4.12:** Selecting a font for installation.

You can install each font separately by clicking on the font name.

# REMOVING TRUETYPE FONTS

Believe it or not, there may come a time when you want to remove some of the fonts on your system. Fonts do take up quite a bit of storage space, and you may find that you need the extra space more than you need lots of fonts. Periodically, you may want to remove those fonts that you do not use or use rarely.

## REMOVING WINDOWS TRUETYPE FONTS

You have two choices for removing Windows TrueType fonts. You can choose to remove the fonts from the fonts list, but still retain the fonts on your hard disk. You also have the option of removing the fonts from your hard disk.

If you remove a font and later work on a document that contained the removed font, Windows substitutes the closest matching font. The name of the original font is displayed in the application's font list so that you can reinstall the font if you choose.

To remove a Windows TrueType font, follow these steps:

1. In the Control Panel, double-click on the Fonts icon.

   The Fonts dialog box is displayed.

2. Select the fonts you want to remove.

   If you want to selectively remove fonts, click on the font name in the List of Fonts list box. You can select more than one font by holding down Ctrl and clicking on the individual fonts that you want to remove. If you want to remove every font, scroll to the bottom of the font list, press Shift and click the right mouse button.

3. Choose the Remove button.

   The Remove Font dialog box is displayed (see fig. 4.13).

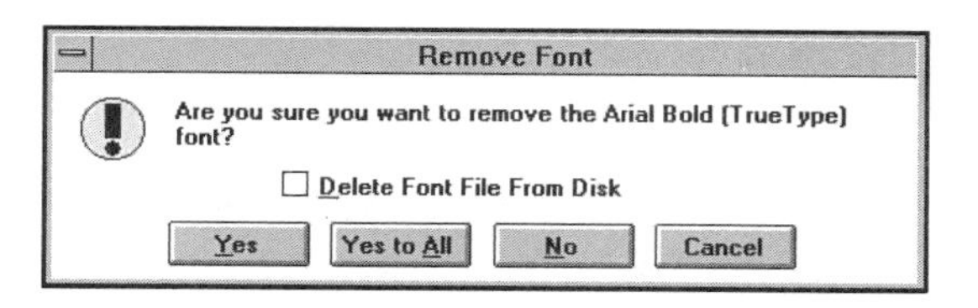


**Figure 4.13:** The Remove Font dialog box.

4. Choose whether you want the fonts deleted from the list or deleted entirely from your hard disk.

   To remove the selected font(s) but retain the font on your system, choose the Yes To All button. You are asked to confirm each deletion. If you want the font(s) removed from your hard disk, click on the Delete Font File From Disk check box and then choose Yes or Yes to All button. An X appears in the Delete Font File From Disk check box.

5. Choose the Close button.

### The Coach Says...

Do not remove the MS Sans Serif screen font. This font is the Windows 3.1 font that is used on-screen in dialog boxes and menus. If you remove this font, you must reinstall Windows to correct the problem.

## REMOVING TRUETYPE MACINTOSH FONTS

Removing a TrueType Macintosh font is very simple. Just drag the font's icon from the System folder to the Trash Can, then empty the Trash Can. The font is deleted.

# INSTANT REPLAY

This chapter discussed:

INSTANT REPLAY

- ✔ The birth of TrueType
- ✔ What is TrueType?
- ✔ Hinting
- ✔ Installing TrueType on Windows and Macintosh
- ✔ Removing TrueType fonts from Windows and Macintosh

# UNDERSTANDING POSTSCRIPT TYPE 1

If any one technology is responsible for the desktop publishing revolution, it's Adobe's PostScript. The publishing world, however, has been in a constant state of change ever since the invention of the printing press.

In this chapter, you learn the following:

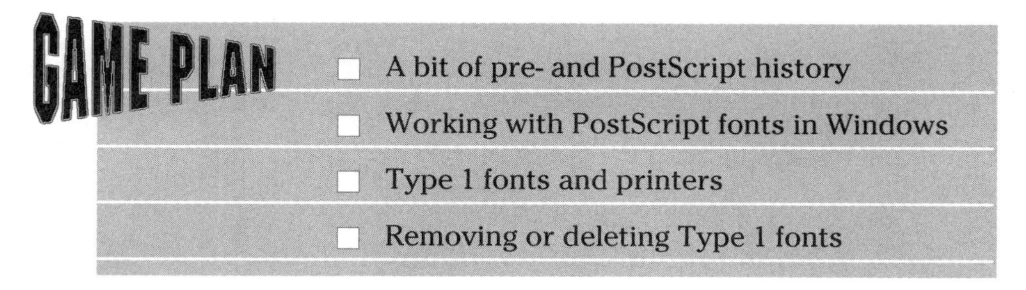


- A bit of pre- and PostScript history
- Working with PostScript fonts in Windows
- Type 1 fonts and printers
- Removing or deleting Type 1 fonts

# FROM GUTENBERG TO THE MASSES

As you learned in Chapter 2, Johannes Gutenberg invented movable type in the mid 1450s. Prior to Gutenberg's innovation, all books were made by hand. Each line of type was painstakingly hand-drawn by a scribe. As you can imagine, this was a tedious process, so printed matter was not the common sight that we all take for granted today. Consequently, books were not intended for, nor enjoyed by, the masses. The printing press enabled the machine production of books. It put many a scribe out of work and helped to begin the information revolution.

Gutenberg's methodology made use of 300 letter typefaces, which, in addition to the common alphanumeric characters, contained ligatures (combination characters, such as ff or ffi, joined as one character) as well as abbreviations. Typographical standards of the day dictated that he follow the look of the scribes, with justified columns. This was made possible by Gutenberg's intricately designed character sets and spacing. Thankfully, today's page layout software handles full justification with aplomb.

### The Coach Says...

The concepts and practices developed by Gutenberg were viewed skeptically by the scribe establishment, but they eventually were accepted. This foreshadows what has happened over the past ten years or so to the printing and publishing industries. Prior to the relatively new upheaval in printing technology, individuals and firms were, in fact, still bound to expensive, painstaking manual or proprietary processes. Until the advent of PostScript, the revolution that Gutenberg began over 500 years ago had not yet reached full force.

## A BLINDING DECADE

The roots of the desktop publishing revolution can be traced back to three men and the companies they founded: Steve Jobs, John Warnock, and Paul Brainerd, of Apple Computer (now of NeXT), Adobe Systems, and the Aldus Corporation, respectively. These three visionaries helped print communication to finally reach the masses. All have made their mark in the past ten years, and in fact, only Apple predates the 1980s (Adobe was founded in 1982 and Aldus in 1984).

It is not an exaggeration to say that without the Apple Macintosh and LaserWriter printer, fewer magazines would be on the newsstands, fewer catalogs would be in your mailbox, and far less would be printed altogether. The Macintosh, and the graphical user interface it popularized, has helped to usher in a new age of information delivery. While the Mac's interface owes plenty to ancient Egyptian hieroglyphics—not to mention Xerox's Palo Alto Research Center (PARC)—it clearly foreshadowed the future.

Without a niche, and the graphic design, printing, and publishing industries, the Mac—and Windows—would not enjoy the popularity that they do today. A slick interface isn't a raison d'être. PostScript and the application programs that use it gave the Mac a reason for being.

## SO WHAT IS POSTSCRIPT, ANYWAY?

Adobe PostScript is a page description language (PDL), enabling the computer to tell the printer how to image (print) a piece of paper. In short, the computer sends a file down to the printer, and the printer's Raster Image Processor (RIP) interprets the file, rasterizes it, and sends the information to be imaged on the page.

The PostScript file sent to the printer is an actual computer program—written in the PostScript language and composed of alphanumeric characters—that actually is readable to the (trained) human eye. The information sent from the RIP is binary. This rasterized, binary data makes up a matrix that defines the page spot-by-spot, enabling users to integrate text and graphics seamlessly and without any technical knowledge of PostScript whatsoever.

The RIP actually sends a bit map down to the printer. But until the RIP has done its work, the file is in vector form (except for bit maps, such as scanned photographs). PostScript describes a page mathematically by constructing and placing paths, which may be text or geometric figures. Mathematical innovation, less than 200 years after Gutenberg, plays a strong role in current computer graphics and publishing systems.

**The Coach Says...**

The work published by mathematician René Descartes, in his *Discours de la méthode* (1637), is a cornerstone of the PostScript language. Within a PostScript page, paths are placed by Cartesian coordinates. Without Descartes' definitions of x and y coordinates, John Warnock would not have had the basis to build a page description language.

## ENOUGH HISTORY, ALREADY! WHAT ABOUT TODAY?

How do people get their information into PostScript, and then onto the printed page? Via the application program! In addition to creating the page description environment, Adobe Systems also has created some of the most powerful application programs. In particular, Adobe Illustrator, throughout its incarnations, has

pushed graphic design and illustration into new realms. Illustrator enables artists to create images quickly and with great precision.

The desktop publishing explosion took place due to the arrival of a number of powerful page layout programs. Aldus PageMaker was the first page layout program to seriously exploit the combination of PostScript and the Mac. Paul Brainerd's company faced an uphill battle, however. Shortsighted professional typographers viewed desktop publishing in general (and the Mac in particular) as a bad joke. In the early days, the Mac was seriously underpowered and had only a handful of fonts. It took plenty of evangelism and hard work to get things rolling.

Just as in Gutenberg's day, the scribe establishment was not amused. While the status quo crowd kept their heads in the sand, however, farsighted publishers took heed of what the Mac heralded: a full-blown page layout workstation that interfaced with word-processed text and graphics for a fraction of the price of existing proprietary systems. An industry was born. PageMaker and similar programs, such as Quark Xpress, enable users to lay out pages as if they are composing pages on a drawing board—a familiar metaphor that has quickly helped computer neophytes become zealots.

### The Coach Says...

PageMaker was the first Mac-based DTP program to make the jump to the Windows platform. As PostScript font libraries arrived on the Mac scene, they too made their way to Windows. Now the big four DTP programs—Aldus PageMaker, Quark Xpress, Adobe Illustrator, and Aldus FreeHand—have migrated to the Windows platform.

# WORKING WITH POSTSCRIPT FONTS IN WINDOWS

Using PostScript fonts in Windows is easy, but not quite as effortless as using TrueType fonts. The big difference is in how you install and manage your fonts. After you are working in your application program—whether it's Word for Windows or PageMaker or whatever—you won't be able to discern any difference.

**The Coach Says...**

To effectively use PostScript fonts in Windows, you must have Adobe Type Manager (ATM) installed. If you are using Windows 3.1, you need ATM Version 2.0 or newer. If you are not sure what version you have installed, take a look at the top left corner of the ATM Control Panel.

## ADOBE TYPE MANAGER

Adobe Type Manager, like all Windows programs, installs itself via a handy Install utility. While it's hardly worth giving you instructions on installing ATM here, a few caveats exist that you need to be aware of.

ATM creates two directories on your hard drive, C:\PSFONTS and C:\PSFONTS\PFM. These two directories are for storing the PostScript outline fonts (PFBs) and the font metrics files (PFMs), respectively.

Each font has two different files: PFB files and PFM files. The PFB files are the actual outline fonts; they are what prints on your printer and displays on your screen. The PFM files are what controls character spacing within your application programs; they include character widths and kerning information. By using a separate file for font metrics, Type 1 fonts can have their kerning tables altered without altering the actual font file.

The ATM Control Panel (see fig. 5.1) displays a list of your installed ATM fonts as well as your ATM settings.

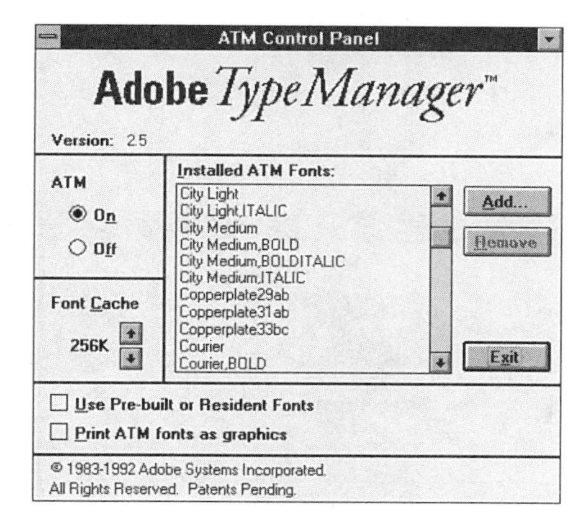


**Figure 5.1:**
Adobe Type Manager Control Panel.

You take charge of your PostScript fonts in the ATM Control Panel and can perform the following tasks:

- ★ Add fonts
- ★ Remove fonts
- ★ Turn ATM on or off
- ★ Adjust the size of the font cache
- ★ Use prebuilt or resident fonts
- ★ Print ATM fonts as graphics

## INSTALLING TYPE 1 FONTS

Type 1 fonts are installed onto your system via the Adobe Type Manager Control Panel. The ATM Control panel usually is easy to find. By default, the ATM Installer places the ATM Control Panel into the Main Window, but you have the option of dragging it to any window you'd like. To display the ATM Control Panel, double-click on it.

Adding fonts with the Adobe Type Manager Control Panel is a straightforward affair. To add a font (or a number of fonts) perform these steps:

1. Click on the Add button to bring up the Add ATM Fonts dialog box (see fig. 5.2).

**Figure 5.2:**

Adding Adobe PostScript Fonts from the New Faces Collection

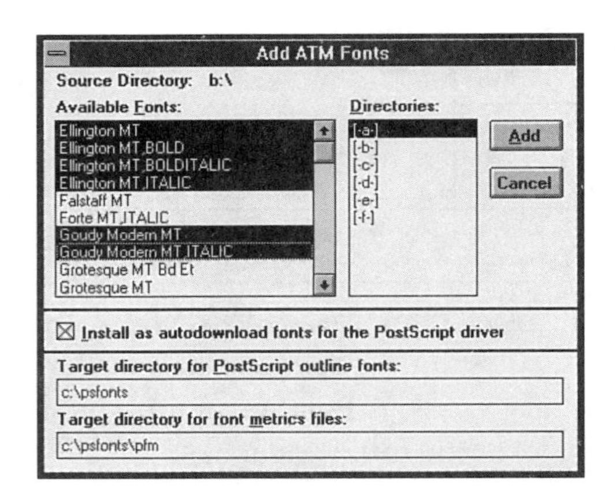


2. Select the disk drive and directory in which the new fonts reside.

3. After you select the fonts you want to add, click the **A**dd button. The cursor changes into an hourglass icon while ATM loads the fonts.

4. After the fonts are loaded (the hourglass icon changes back to your normal cursor), you can continue to add more fonts from other disks/directories.
5. When you are done loading fonts, click on Exit.

Prior to ATM Version 2.5, whenever you added PostScript fonts to your system with ATM, you were required to restart Windows to be able to use the fonts. Version 2.5, released in late 1992, enables you to immediately begin using the fonts. Certain applications, such as Aldus PageMaker 4.0 and Lotus AmiPro 3.0, need to take another look at the printer setup; consequently, you need to reselect your printer for the changes to take effect. Other programs may need to be closed down and relaunched.

### The Coach Says...

Some foundries place the PFM and PFB files in different directories on their distribution disks. This can be a hassle when you are installing fonts, particularly if you don't know where the files are stored. Fear not. You can Alt-Tab out to the Windows File Manager (if it is already running; if not, go to the Program Manager and launch the File Manager) to snoop around on the distribution disk. After you've found the right directory, Alt-Tab back to the ATM Control Panel.

To make things easier on yourself, you always can create a dummy directory on your hard drive to use as an intermediate step when loading fonts. Just copy the distribution PFM and PFB files to the dummy directory with the Windows File Manager. Then, when you install fonts with ATM, you have to look in only one directory. Don't forget to clean up this directory when you're done!

## TWEAKING ATM

As you can see by looking at the ATM Control Panel, a number of ATM settings can be altered. You may want to make changes here, depending on your system configuration. You should, for example, adjust the size of the font cache in relation to the amount of RAM you have in your machine.

### The Coach Says...

The *font cache* is a place in memory that ATM builds for rasterized screen font storage. While the default value is set to 96K, Adobe recommends that you allocate no more than 64K times the number of megabytes of RAM in your system (#RAM * 64k = Size of font cache).

This means that if you have 4M of RAM, you should set the font cache for no higher than 256K; 8M of RAM no higher than 512K; and so on. You definitely can get away with a more modest setting. Use the higher setting if you are a "fontaholic"; use the lower setting if you use a limited number of fonts. In other words, the more fonts you use in a document at one time, the larger your font cache should be.

Check the Use Pre-built or Resident Font option if you want to use PCL fonts. If you are not sure whether you have any prebuilt fonts installed, go to the Program Manager Main Windows and double-click on Control Panel. (This is different than the ATM Control Panel.) After you bring up the Fonts dialog box (by double-clicking on its icon), you can check to see if any prebuilt fonts are installed. If you have an HP LaserJet or IBM Lexmark printer, and are running Windows 3.1, ATM 2.5 can create soft fonts "on-the-fly." If so, you can take advantage of ATM, and turn the Pre-built option off.

If you select the Print ATM Fonts As Graphics option, ATM alters the files sent down to the printer by converting all text to graphics. If, however, you have set up your Windows print driver with the Print TrueType Fonts as Graphics option enabled, this overrides the ATM panel setting and all ATM fonts also print as graphics.

## TYPE 1 FONTS AND PRINTERS

It's important to know what kind of printer you have, and especially whether it is a PostScript printer or not. All printers—and printer drivers—are not created equal.

> **The Coach Says...**
>
> A *printer driver* is a small program that allows other programs (like Windows) to communicate with your printer. The printer driver has a great effect on output—things like available resident fonts, page sizes, margins, and the like. While you can use a printer driver with a printer other than the one it was written for, you get the best results by using the proper printer driver.

On a PostScript printer, Type 1 fonts can be either printer-resident or downloaded—either with the job or before you send a job. Once again, the right printer driver is important; otherwise, Windows does not know which fonts are resident (or built-in) on your printer, and hence, those fonts may not show up on the font menus of the various application programs.

## BASE 13 OR BASE 35? WHAT IS THIS, A MATH QUIZ?

As you may recall, the Apple LaserWriter was the first PostScript laser printer. It came with what was known as the *Base 13* fonts. We're not talking about an oddball number system here, but the number of printer resident fonts. If you see a reference to the Base 13 font load, you should know that it consists of the following font families and the Symbol font:

- ★ Courier
- ★ Helvetica
- ★ Times

These 13 fonts are in the read-only memory (ROM) of all PostScript output devices.

Later generations of the LaserWriter, as well as PostScript printers from a number of manufacturers, come with what is known as the *Base 35* font load. In addition to the Base 13 fonts, the Base 35 fonts include the following font families:

- ★ Avant Garde
- ★ Bookman
- ★ Helvetica Narrow
- ★ New Century Schoolbook
- ★ Palatino
- ★ Zapf Chancery Medium Italic
- ★ Zapf Dingbats

**The Coach Says...**

Some PostScript printers—such as the newer offerings from QMS—may come with more than just the Base 35 fonts.

Printers do not always come with floppy disk versions of their printer resident fonts. Fear not, you can use the fonts on the printer, but you won't have the pretty screen display that ATM and the PFB files provide. That's why Adobe sells its Plus Pack. Plus Pack provides the PFB files you need for the proper screen representation. If you don't have the PFB for a printer-resident font, ATM will dummy in either Helvetica or Times, depending on whether the font is of a serif or sans serif design.

**The Coach Says...**

When installing the Adobe Plus Pack fonts with ATM 2.5, make sure that the Install as autodownload fonts for the PostScript Driver option is not selected. Otherwise, you'll be needlessly downloading those printer-resident fonts.

## WHAT IS DOWNLOADING?

You've probably run across the term *downloading*, and are thinking, "Vas is das?". Downloading is the process of sending a font down to the printer. You need to download any PostScript font you use, in any particular document, with the exceptions of your printer's base font load. This is not as complicated as you may think. After things are set up correctly, you shouldn't have to consciously download fonts; the Windows PostScript printer driver does it for you. The trick, however, is setting up your system correctly.

When you load PostScript fonts with the ATM Control Panel, ATM adds lines to the `[PostScript,portname]` section of your WIN.INI file. These lines tell the printer driver where to look for the font metrics (PFM) and font outline (PFB) files. Each line looks something like the following:

```
softfont1=c:\psfonts\pfm\gdrg____.pfm;c:\psfonts\gdrg____.pfb
```

What does this line mean? It's really not that tough to understand. Take another look. The `softfont#` entry is just that. In this case, it's the first soft font that is installed.

The next section in the line g code, up to the semicolon, is the path and file name of the PFM file. The last section (everything after the semicolon) is the path and file name of the PFB file.

With an entry like that, the Windows PostScript printer driver automatically downloads the font—in this case, Adobe Garamond—each time it sends a job that uses that font down to the printer. If you don't want to download the font (perhaps you already manually downloaded the font), you can make a simple change to the line so that it reads like this:

```
softfont1=c:\psfonts\pfm\gdrg____.pfm
```

If you don't want a font to autodownload, all you need to do is remove everything in the reference line after the semicolon. But this does not mean that you need to manually edit the WIN.INI file. Adobe Type Manager Version 2.5 will edit WIN.INI for you—assuming that you correctly install your fonts, that is. If you try this and find that your printer is coughing up Courier (typewriter) type where you expected other fonts, those other fonts are not resident.

### The Coach Says...

The Mac has a big advantage over Windows when it comes to font handling. Specifically, the Mac has the capability to query the printer to see if a particular font is already resident at the printer. This is possible because the Mac and its printer maintain a two-way conversation over a LocalTalk wire. In contrast, the typical Windows PC just throws files—most commonly, down a parallel cable—at the printer, blindly downloading fonts with each job (unless you do some WIN.INI editing).

It's your responsibility to determine how you want to handle downloading fonts. The no-brainer approach is to let Windows download everything except for the Base fonts. This can waste precious minutes at print time, but you won't be looking at Courier, either.

You can make your life far easier with a hard drive-equipped PostScript printer. If you are in the market for a new PostScript printer and you use a large variety of fonts, consider a printer with a SCSI (pronounced "scuzzy") port. This enables you to attach a hard drive filled with printer fonts. This way, you download each font only once—to the printer's hard drive—and save untold hours at print time.

**The Coach Says...**

The typical Mac office shares printers via LocalTalk wiring. If you are a PC user, don't think that you can't join the party! You can hook your PC into the LocalTalk network by adding a LocalTalk card. These cards are available from a number of manufacturers, including DayStar and Farallon.

## MANUALLY DOWNLOADING FONTS

If your printer has plenty of RAM (random-access memory), and you consistently use a number of downloadable PostScript fonts, you may want to consider manually downloading those specific fonts.

You can download fonts with a variety of font downloading utility programs. If you've purchased additional Adobe fonts, you'll find a pair of font downloader utilities on the distribution disks. You'll want to use the PCSEND utility if your printer is attached via a parallel port (such as LPT1), and the PSDOWN utility if your

printer is connected to a serial port (COM1, for instance). A number of other notable downloaders are available. WinPSX, for example, is a slick little freeware Windows PostScript font downloader that once you see, you won't want to be without.

The amount of RAM installed in your printer determines how many fonts can be downloaded. If your work is routinely mega-fonts-per-page, consider installing more printer RAM. How do you know if you need more printer RAM? Your printer will substitute fonts or balk on pages with a number of fonts. Downloading strategies vary, but all will go easier with more RAM. If you use the same fonts all day long, you may want to build a routine into your AUTOEXEC.BAT file that automatically downloads those fonts upon system boot-up.

## REMOVING OR DELETING TYPE 1 FONTS

Sometimes you may want to either remove or delete PostScript fonts from your system. What is the difference? Removing fonts simply takes those fonts off the Available ATM Fonts list, while deleting fonts does that in addition to deleting the font files off of your system's hard disk. In most cases, you will just want to remove fonts, rather than delete them (see fig. 5.3).

Removing ATM fonts.

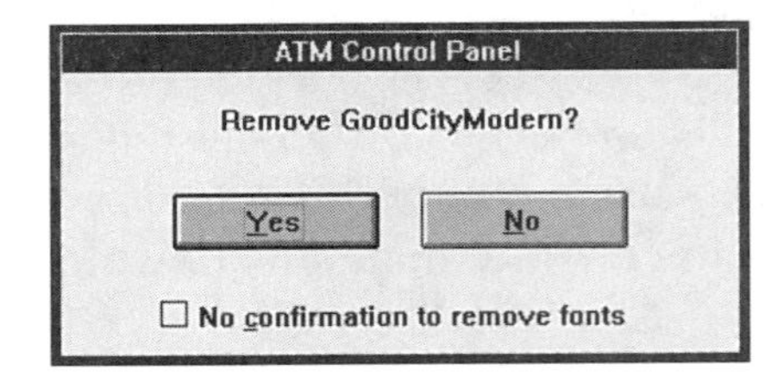


To remove fonts, use the ATM Control Panel. To delete fonts, you must physically delete them from the \PSFONTS and \PSFONTS\PFM directories, after they have been removed from

ATM. This can be a pain in the neck. If you are not sure which font is which file, you will have to take a careful peek into the ATM.INI file with Windows Notepad (or another text editor).

**The Coach Says...**

If you have hundreds—or even thousands—of fonts, and find yourself wrestling with them on a daily basis, you'll want to beef up your font handling tools. Ares FontMinder is an excellent font management utility worthy of your consideration.

## ADVANCED FONT MANAGEMENT WITH ARES FONTMINDER

Need a way to control a burgeoning array of fonts? Ares FontMinder tames the font beast by organizing your fonts into easily installable (and de-installable) font packs. These font packs can contain any array of Type 1 and/or TrueType fonts. (Although each font pack can consist of only either PS or TT fonts, you can have multiple packs installed.) This enables you to arrange your fonts according to clients, recurring jobs, or operators.

The idea is to load only the fonts you need, rather than your entire collection. FontMinder snoops through your system and assembles a master library of fonts (see fig. 5.4). This makes creating font packs a simple task. With an easy way to swap between font loads, your system operates faster, and with less overhead.

As you can see in figure 5.5, most of the action takes place in FontMinder's main window. Moving fonts and font packs around is as easy as dragging them about. Ares even includes a whimsical trash can that opens and slams shut as you remove fonts.

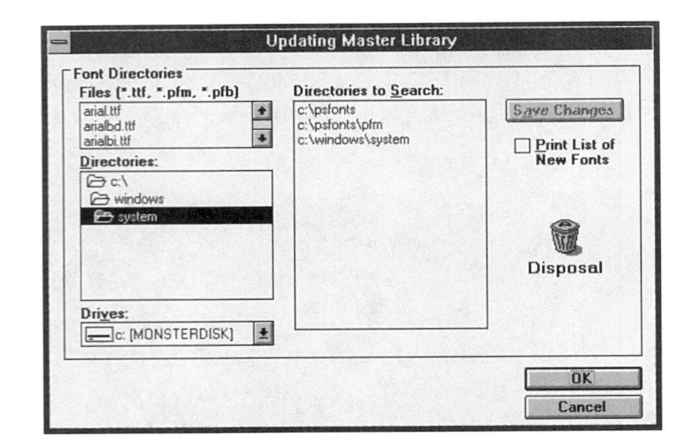


**Figure 5.4:** Updating the Ares FontMinder Master Library.

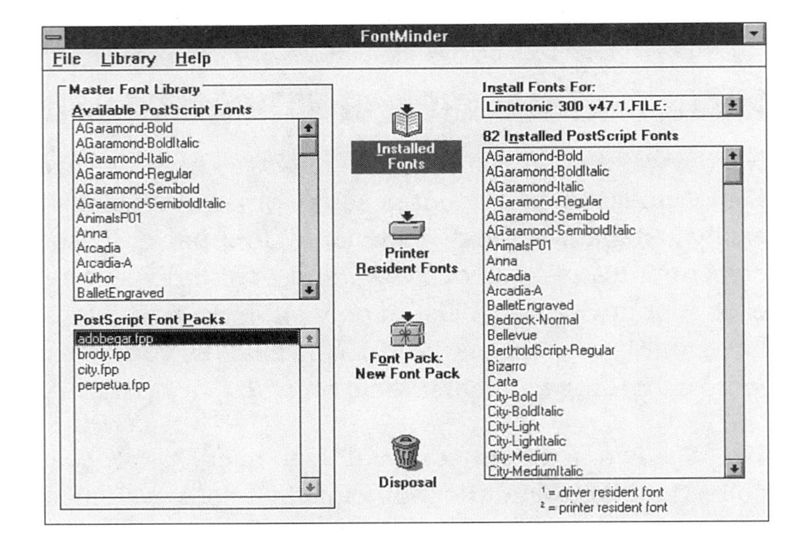


**Figure 5.5:** Taming the font monster with Ares FontMinder.

## The Coach Says...

FontMinder evolved from the successful shareware program, FontManager. Developer Dennis Harrington's pet project became so popular that it attracted the attention of Ares, and has now grown into a full-fledged commercial application worthy of any serious DTPer. The program has received much acclaim from the Windows DTP community.

# INSTANT REPLAY

In this chapter, you learned how PostScript works, and how it can work for you.

- ✔ A bit of pre- and PostScript history
- ✔ Working with PostScript fonts in Windows
- ✔ Type 1 fonts and printers
- ✔ Removing or deleting Type 1 fonts

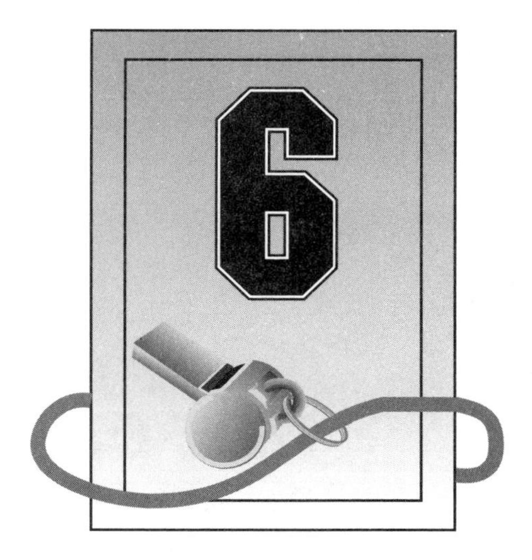


# SHOPPING FOR FONTS

Shopping and acquiring fonts can be fun, but it also can be as confusing as the soda pop aisle of your local supermarket. The choices seem endless! All those varieties make decision-making tough. The same can be said of choosing fonts. What type of fonts should you buy? Which faces are the best choices? How much do fonts cost?

This chapter answers these questions and helps you determine what your needs are. Developing a plan can help simplify this process. After you read this chapter, you should have the information you need to develop a practical font library.

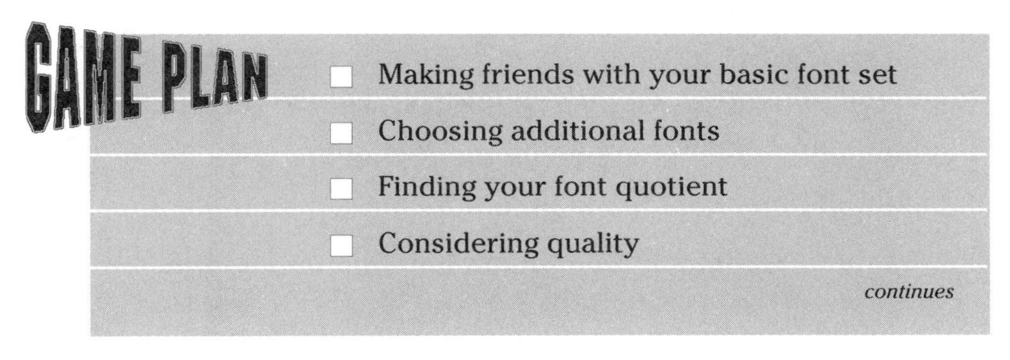

*continued*

- ☐ Considering cost
- ☐ Purchasing fonts

# WHERE DO YOU BEGIN?

If you are not an experienced user of fonts, begin by looking at the fonts that are currently installed on your system. This set of fonts might be all that you need for right now. As you grow more accustomed to using fonts, you always can purchase more faces later.

## STANDARD WINDOWS FONTS

Windows 3.1, for example, comes with 14 TrueType fonts. These 14 fonts are divided into three distinct families and two specialty fonts. The three families include Times New Roman, Arial, and Courier. Within each family, you can choose from roman, italic, bold, and bold italic. The specialty faces are Symbol and Wingdings. These faces include symbols and pictures, such as arrows, boxes, four-leaf clovers, and so on. Figure 6.1 illustrates the Windows 3.1 basic font set.

**The Coach Says...**

You cannot go wrong if you choose Times New Roman for body text. It is a good readable font. Arial, on the other hand, makes an excellent font for headings.

Times New Roman
*Times New Roman Italic*
**Times New Roman Bold**
***Times New Roman Bold Italic***
Arial
*Arial Italic*
**Arial Bold**
***Arial Bold Italic***
Courier
*Courier Italic*
**Courier Bold**
***Courier Bold Italic***
Σψμβολ

**Figure 6.1:** The Windows 3.1 standard font set.

## STANDARD MACINTOSH FONTS

If you are a Macintosh user, you have more standard choices than the Windows 3.1 user. Your basic font set includes 11 font families. These families are Avant Garde, Bookman, Courier, Helvetica, Helvetica Narrow, New Century Schoolbook, Palatino, Symbol, Times, Zapf Chancery, and Zapf Dingbats. Figure 6.2 illustrates these basic font families.

### The Coach Says...

Of the standard Macintosh font set, Times, New Century Schoolbook, Bookman, and Palatino make the best choices if you are typing long passages of body text. Helvetica, Avant Garde, and Narrow Helvetica make the best choices for headings. Use Zapf Chancery, which is an elegant script face, for invitations or certificates.

**Figure 6.2:**

The Macintosh basic font set.

abcdefghijklmnopqrstuvwxyz
Avante Garde

abcdefghijklmnopqrstuvwxyz
Bookman

abcdefghijklmnopqrstuvwxyz
Courier

abcdefghijklmnopqrstuvwxyz
Helvetica

abcdefghijklmnopqrstuvwxyz
Helvetica Narrow

abcdefghijklmnopqrstuvwxyz
New Century Schoolbook

abcdefghijklmnopqrstuvwxyz
Palatino

αβχδεφγηιφκλμνοπθρστυϖωξψζ
Symbol

abcdefghijklmnopqrstuvwxyz
Times

*abcdefghijklmnopqrstuvwxyz*
Zapf Chancery

❀❂✿❄❅❆❇❈❉❊❋●❍■❏❐❑❒▲▼◆❖◗❘❙❚
Zapf Dingbats

## WHAT ABOUT DOS?

Most DOS applications use those fonts that are available in the printer. The fonts can be built into the printer or stored on a font cartridge that you insert in the printer. DOS has lagged behind Windows and the Macintosh because it is a character-based

operating system. The characters that appear on-screen are of a fixed width.

You can use some DOS programs to help you get around this problem. These programs have a graphics mode that enables you to display the page. LaserTools makes an add on for WordPerfect, called PrimeType, that is a special version of Adobe Type Manager. A new font supplier, called SoftMaker, includes a DOS font scaling engine for a number of popular DOS programs, such as WordPerfect, Word, and Works. This font scaler enables you to use PostScript Type 1 fonts with these DOS applications. Other DOS desktop publishing programs provide their own graphics interface on top of DOS. These programs usually use their own proprietary font formats.

**The Coach Says...**

If you are serious about using fonts for serious professional business purposes, however, you probably should be using Windows 3.1 or a Macintosh.

## WHAT COMES NEXT?

Eventually, you will want to add more typefaces to your font library. With so many choices, you might wonder how to begin to choose other fonts. The most important piece of advice is not to be an impulse shopper. Impulse buys can be costly mistakes. Remember the time that you joined the gym only to go once or twice? It seemed like a good idea at the time. The same advice goes for font purchases. Make sure that you can use the font in the documents you create. Decorative fonts can be fun to play with, but may not add any real value to your library.

**The Coach Says...**

You can experiment with fonts without investing a lot of money by using shareware or freeware fonts. Shareware and freeware are software that are developed by an individual and offered to the public for a small fee. You can try a font, and if you like it, you then pay a small fee to the developer. Shareware is covered in more detail later in this chapter.

## FINDING THE RIGHT FONT PACKAGES

As you read magazines, advertisements, or brochures, save or copy those articles that contain faces that attract your attention. Notice the way that professional designers use type. You might want to subscribe to publications that are geared toward the publishing or design businesses. Examples include *Publish*, which is a magazine aimed at people in information design, and *U&lc* (Upper- and Lowercase), which is a tabloid that is written for typographers. Another publication is *EC&I* (Electronic Composition and Imaging), which is published in Canada.

**The Coach Says...**

As you find examples of fonts or designs that you like, make photocopies and file them in a three-ring binder. You also can place in the binder any articles that you find in professional publications that discuss trends in fonts or design tips.

## DETERMINING YOUR FONT USER QUOTIENT

No, you don't need to do mathematical computation to figure out your font quotient. You just need to evaluate what you need to do with fonts and determine which fonts you feel comfortable with. You must like a font to use it successfully. Font styles can be very subjective and personal. You probably have had your hair cut in a popular style only to feel that it doesn't suit you. This analogy works just as well with fonts. If you do not feel comfortable with the look of a font, you will not feel comfortable using it in your documents.

If you are a conservative person, stick with the more classical font styles. Classical fonts include such faces as Times Roman, Garamond, Palatino, and Helvetica. If you are a daring individual, go ahead and buy some of those wild and crazy fonts, such as Reporter, Lithos, or Ironwood. Just remember that trends come and go, but the basic conservative text fonts always remain in style.

You should consider the way in which you use fonts. Are you a typical business user who creates business letters, memos, and occasional reports? Are you responsible for creating visual presentations for a marketing department? Are you a budding desktop publisher who creates newsletters, press releases, and brochures? Are you a home user who enjoys having a few fun faces to jazz up a neighborhood newsletter or a flyer advertising your garage sale?

Tailor your font purchases to your needs. If your business produces technical documentation, for example, you probably don't have a great need for script typefaces. If you produce a wide range of documents, however, you might need a much larger font library.

> **The Coach Says...**
> If your business uses a particular typeface in its company trademark or logo, you probably should invest in that font. Buying this font enhances and continues the tone that your logo or letterhead creates. You essentially establish an entire look and feel for your company.

Table 6.1 offers suggestions for making font purchases. The table is organized into three sections. The first section displays typefaces that make good text faces; the second section displays typefaces that make good display faces; and the third section displays suggested specialty faces. With the hundreds of typefaces from which you can choose, this table is far from complete, but it does provide you with a foundation from which to work.

**Table 6.1: Suggested Typefaces**

| *Text Faces* | |
|---|---|
| Bookman | Bookman is a strong, friendly, and somewhat casual text face. Although it can be difficult to read in long passages of text, it makes an excellent choice for advertisements, brochures, or any document that contains short passages of text. |
| Garamond | Garamond is a good choice of typeface for business documents. You often see Garamond used in legal and financial applications. |
| Goudy | Use Goudy for business documents such as annual reports, letterhead, and business cards. Goudy also looks good in larger point sizes. |

| Text Faces | |
|---|---|
| Century Schoolbook | As the name implies, Century Schoolbook is the perfect choice for books. This face has a high degree of readability. |
| Palatino | Palatino is an elegant text face. This face has an almost classical appeal. You can use this face in a variety of ways from books to brochures. |
| ***Display Typefaces*** | |
| **Helvetica Black** | Helvetica Black is a strong sans serif typeface. This face looks good with any serif text font. It is versatile and prints well. |
| **Poster Bodoni** | Poster Bodoni is characterized by its great variation in stroke weight. Because of this variation, Poster Bodoni can be hard to read. Use it in short headlines when you want a formal look. |
| Futura | Futura is characterized by its large, round, and simple letters. Use futura in headlines when you want a more modern look. |
| Avant Garde | Avant Garde is another modern looking sans serif face. Use it for ads, presentations, and headlines. |
| ***Specialty Typefaces*** | |
| BluePrint | BluePrint is an informal typface with the look of handlettering. Use it when you want your message to be informal. |
| *Brush Script* | Brush Script is an informal script typeface. Use this face for invitations or signs. |

*continues*

**Table 6.1:**
**Continued**

| *Specialty Typefaces* | |
| --- | --- |
| Fette Fraktur | Fette Fraktur is a bold gothic typeface. This typeface is perfect for certificates and awards. |
| *Zapf Chancery* | Zapf Chancery is an elegant script face that is perfect for formal invitations. |

## A FONT BY ANY OTHER NAME

Note that font names are protected under copyright laws, but the artwork or actual style of the face is not. As you thumb through type catalogs from different manufacturers, you may notice that some fonts appear identical but have different names. Do not purchase fonts by name alone. Always compare fonts so that you are sure that you are ordering the face that you want.

### The Coach Says...

The reasoning behind the copyright laws is rather confusing. The court decided that the alphabet cannot be copyrighted. Although artwork can be copyrighted, letter forms cannot be copyrighted. *Letter forms* in computer fonts are made up of pixel patterns, and pixels are not unique in the computer industry. A *pixel* is the smallest element that your computer screen can show. Each letter is made up of many pixels. The court decision handed down, therefore, states that a specific font pixel pattern cannot be copyrighted.

Table 6.2 provides reconized font names along with faces that look very similar, but that have different names. The traditional Caslon

face, for example, has competition from a similar face called Casper. Often the names are very similar. Although this table is far from complete, it does give you an idea of the number of clone faces that exist.

**Table 6.2: Alternate Font Names**

| *Font Name* | *Alternate Name(s)* |
|---|---|
| Americana | America |
| Antique Olive | Olive Antique, Provence |
| Architect | BluePrint |
| Bauhaus | Bahamas |
| Bembo | Ambo |
| Bodoni | Bodnoff, Galileo |
| Bookman | Brooklyn |
| Broadway | Ritz |
| Brush Script | Banff |
| Caslon | Casper |
| Cooper Black | Cupertino |
| Eras | Erie |
| Fette Fraktur | Frankenstein |
| Futura | Modern |
| Helvetica | Arial, Switzerland |
| Hobo | Hoboken |
| Machine | Computer |
| Mistral | Mystical, Staccato 222 |

*continues*

**Table 6.2: Continued**

| *Font Name* | *Alternate Name(s)* |
|---|---|
| Neon | Quantum |
| Palatino | Palladia |
| Peignot | Penguin, Peigno |
| Stencil | Stamp |
| Tiffany | Timpani |
| Times Roman | Times New Roman |
| Univers | Univers (WN) |
| Zapf Chancery | Zurich Calligraphic |

**The Coach Says...**

Many font companies offer free type catalogs or posters. Appendix A is a compilation of font suppliers complete with addresses and phone numbers. Contact these suppliers for free literature and samples of the fonts that they offer.

## QUALITY QUESTIONS

Before you purchase any font, you should consider the output quality that you need. If you are not planning on having your documents professionally printed, you probably need to purchase faces that print well at lower printer resolutions, such as 300dpi.

Printer resolution is the density of the printed type. Most laser printers print at 300 dpi; professional typesetters at 1200 to 3600 dpi. The higher the number, the better the printed quality. If you

are producing documents on a typical laser printer and then photocopying the documents for mass distribution, you must consider each typeface you buy. The process of photocopying can thicken the individual characters and distort the face. If you set the photocopy machine for light copies, you may lose delicate serifs.

Always request a sample of the typeface you want to purchase printed in several point sizes at the resolution you intend to print. Point size can greatly affect resolution. Some faces contain jagged edges and show character flaws if printed at large point sizes. If printed at small point sizes (under 8 point), the faces may become fuzzy.

### The Coach Says...

*Dots-per-inch (dpi)* is the measurement of the screen and printer resolution. Dots-per-inch is the measurement of the number of dots that the device produces per linear inch. The higher the dpi, the better quality you see in resolution.

The following guidelines can help you determine whether a font is a good choice:

- ★ Look at the entire printed alphabet. The letters should appear even. Even letters are those in which ascending and descending strokes appear of equal size across the entire character set.
- ★ Characters should appear of the same weight.
- ★ Look at the space between letters, which is known as letterspacing. Loose letterspacing may make documents harder to read.

> **The Coach Says...**
> Fonts that have been developed since the advent of scalable font technology, such as Lucida and Stone, print clearly at low printer resolutions. Fonts that were developed before this time do not print as clearly at lower resolutions.

# WHERE TO BUY FONTS

You have many options when it comes to purchasing fonts. Suppliers range from major manufacturers, mail-order houses, shareware sources, and even freeware. The price you pay can range from expensive to moderate to free.

## BUYING FROM THE MANUFACTURER

When you buy from a manufacturer, you purchase directly or buy from a retail outlet. Fonts usually are sold in font families for $100 to $300 each. A font family consists of the roman, italic, bold, and bold italic of a particular face, such as Times Roman.

> **The Coach Says...**
> Although these prices may seem high to you, remember that the cost also includes expert technical assistance. If you have installation problems or printing problems, technical assistance can be a lifesaver.

The following list includes the major font manufacturers:

★ **Adobe Systems.** Adobe has the most inclusive type library of any manufacturer. With over 1,000 typefaces, its library of fonts includes the most basic text faces to typefaces designed for specific applications, such as flyers or business letters. Prices range from $100 to $400.

  Adobe designs its own fonts, but also licenses fonts from other manufacturers. Adobe does not supply fonts in TrueType format.

★ **Agfa Compugraphic.** Agfa's library of fonts includes nearly 200 different faces. You can purchase the fonts in the Intellifont format, which are supported by the Hewlett-Packard LaserJet III and LaserJet IIIP printers. You also can purchase these fonts in PostScript Type 1 format. Agfa also has a limited number of TrueType faces available. Prices vary. A basic package including one typeface family costs $99. A package that includes 12 fonts costs $159.

  The Intellifont faces purchased from Agfa include Type Director, which is a type manager utility that enables you to convert the basic outlines of letters in the font so that you can use them in applications such as Windows, Microsoft Word, WordPerfect for Windows, and Ventura Publisher.

★ **Bitstream.** Bitstream originally entered the market with bit-mapped fonts for DOS applications such as WordPerfect. This file format is denoted by the .BCO file extension. Each application's font installation routine takes this format and creates a screen and printer font. You can choose from 1,100 typefaces in this original format. Prices range from $129 for four typefaces to $199 for 12 typefaces.

Bitstream's new format is called Speedo. The advantage to Speedo is that its character sets can contain up to 560 characters and symbols. Bitstream also provides over 1,000 Type 1 PostScript fonts for the Macintosh. They have converted these fonts to PC format and are releasing them now. Bitstream plans to release TrueType fonts in the future.

★ **Casady & Greene.** This company does not offer the extensive collections of other manufacturers, but it does offer unique faces. These faces include Cyrillic and Eastern European faces. You can purchase Type 1 or TrueType fonts from $99 to $140.

★ **Digi-Fonts.** Digi-Fonts offers a wide library of fonts that are based on original typefaces from other sources. A quick look at a type catalog should show you that many of their faces are designs of standard faces. Digi-Font's Frisco, for example, is similar to Futura. This company also offers foreign language fonts in Hebrew, Greek, Cyrillic, Gujarati, and Hindi.

The library contains 400 faces that are available in Intellifont and PCL 5 formats. Digi-Fonts plans to develop a Type 1 font library in the near future.

★ **Digital Typeface.** This manufacturer has licensed most of its 100 typefaces from the URW type foundry of Germany. These fonts represent some of the most recognized faces in the typographic industry. Currently, you can purchase these fonts in Type 1 format. The company plans to release 500 more faces soon, with TrueType and Nimbus/Q format support.

★ **Hewlett-Packard.** This company sells font cartridges that contain 24 to 26 fonts. Working in conjunction with Agfa Compugraphics, this company developed the PCL 5 format used on HP LaserJet III and III P printers.

★ **Image Club Graphics.** Some of the most unusual fonts can be purchased from Image Club Graphics. This company specializes in display faces that include letters constructed from paper clips to letters shaded in gray scale. Fonts are offered in Type 1 and Type 3 formats.

Prices are extremely reasonable, ranging from $100 for a font family of four faces to $89 for various starter packs that include 10 fonts. Recently Image Club has dropped prices, and you can obtain some faces as low as $15 with a $50 minimum order.

★ **Linotype-Hell.** This type manufacturer is a merger of two companies. Linotype was the original manufacturer of the linotype machine that was used in the hot-metal days of typesetting. Hell is a scanner manufacturer. The two companies joined forces to create some of the most respected high-quality fonts available today.

The company offers over 2,000 faces in its proprietary format for Linotype typesetters. The company plans to translate all 2,000 faces into Type 1 format. You can purchase nearly 1,000 Type 1 faces currently directly through Linotype-Hell or through Adobe. Packages cost from $100 to $400.

★ **Microsoft Corporation.** Microsoft introduced its TrueType Font Pack for Windows 3.1, which included 44 scalable fonts. Bigelow & Holmes supplied the 22 Lucida fonts, and Monotype Typography, Inc. supplied the other 22 fonts, which range from Century Schoolbook to Corsiva. Note that most PostScript printers contain this standard set of fonts.

★ **Monotype Typography, Inc.** One of the oldest companies, Monotype offers nearly 400 fonts in Type 1 format. Packages can be purchased for $100 to $300. TrueType font packages also are available, and availabilty will

continue to expand as this foundry releases its complete line in this format.

Many of Monotype's fonts are designed for high-quality printing on professional, high-resolution typesetters. If quality is what you need, this manufacturer can supply the font you want.

★ **URW.** This company is a well-known type foundry in Germany. It produces fonts in the Nimbus/Q format. This manufacturer supplies fonts for programs such as SuperPrint from Zenographics. This company plans to release their font library to the public soon. Until this time, its fonts were licensed to other font suppliers.

## PURCHASING FONTS ON CDS

If you happen to own a CD-ROM drive, purchasing fonts can be convenient and cost-efficient. Many font manufacturers are offering their complete font libraries on CD-ROM. You purchase the CD, which contains the complete library, for around $100. For this price you receive a specified number of start-up fonts. The remaining fonts are locked from your use. These fonts remain locked, or encrypted, until you pay a fee for them.

Most manufacturers provide either printed samples or on-line screen display of the fonts that you can purchase. After you decide which fonts you want to purchase, you call the manufacturer. You then supply your user code and credit card number, and the manufacturer then provides an access code so that you can unlock the fonts you ordered. After you have unlocked the printer outlines, you can transfer them to your hard disk.

## ADVANTAGES AND DISADVANTAGES

This method of purchase can be convenient for several reasons. You can buy only those faces that you really need. You no longer need to buy a complete font family of four faces. Purchasing in this manner can be less expensive. You only pay for those faces you are going to use. You can have the face you purchase almost immediately. If you need a typeface fast, this method cannot be beat.

### The Coach Says...

This method also has some drawbacks. The typeface you purchase is registered to one computer only. Make a copy of your access codes in case your system crashes and you find that you must reopen the fonts that you previously purchased. Don't fall into making impulse purchases. You should consider carefully each font purchase you make.

Table 6.3 illustrates some manufacturers who currently offer their font collections on CD-ROM.

**Table 6.3:**
**Font Collections on CD-ROM**

**Product Name:** Adobe Type on Call 1.0

**Manufacturer:** Adobe Systems

**Price:** $99

**Description:** Disc includes 1,350 faces; 13 available at purchase with another 13 upon registration. Cost per style is $50. All faces are in the Type 1 format and are available for PC and Macintosh.

*continues*

**Table 6.3: Continued**

**Product Name:** Agfatype Collection 3.0

**Manufacturer:** Agfa Compugraphic

**Price:** $99

**Description:** Disc includes 1,900 faces; one face at purchase with an additional 20 faces plus one Pi and Symbol face at registration. Each additional face costs $50. Both TrueType and Type 1 versions are available for the Macintosh.

**Product Name:** Bitstream Type Treasury 1.0

**Manufacturer:** Bitstream Inc.

**Price:** $69

**Description:** Disc includes 1,030 faces; six faces upon registration. You can buy an additional one to three faces for $39 each, 4 to 10 faces for $30 each, and over 11 faces for $25 each. You can purchase both TrueType and Type 1 formats for the Macintosh.

**Product Name:** Castcraft Optifont CD Series Volume 1

**Manufacturer:** Castcraft Software, Inc.

**Price:** $996

**Description:** This disc includes 400 fonts that are all unlocked and ready to use. Available font formats include Type 1, Type 3, and TrueType for both PC and Macintosh.

**Product Name:** The Font Company Compact Disc Type Library 2.0

**Manufacturer:** The Font Company

**Price:** $29.95

**Description:** This disc includes 1,653 fonts. You receive one font at registration. Prices per additional font are $39.95 each, or $34.95 for two to five additional fonts; $29.95 for six to nine additional fonts, and $24.95 for 10 or more additional fonts. This disc includes Type 1 fonts for the Macintosh.

**Product:** Image Club Graphics Letterpress 2.0

**Manufacturer:** Image Club Graphics

**Price:** $3,999

**Description:** You receive 626 unlocked fonts on this disc. You can order the disc in Type 1 format or Type 3 format for both the PC and Macintosh.

**Product:** Monotype Typography Fonefonts 92.3

**Manufacturer:** Monotype

**Price:** $49.99

**Description:** This disc includes 1,546 fonts, and you receive eight fonts at registration. To order more fonts you pay $40 each for each additional one to three fonts; $35 for each additional four to eight fonts; $30 each for each nine to 99 fonts; and $25 for ordering over 100 fonts. You can purchase Type 1 fonts for PC and Macintosh.

As this technology grows and more people own CD-ROM drives, additional manufacturers will release their type libraries on CD. If you are interested in obtaining fonts on CD, consult Appendix A. Most manufacturers have toll-free phone numbers.

## SHAREWARE

Another source for fonts is shareware. *Shareware* is software that is developed by an individual or an organization and then offered

to the public for a small fee. You can try the software, and if you decide that you want to keep it, you pay a registration fee to the author of the software.

You can find many shareware TrueType fonts that are reasonable in price. Some are clones of traditional faces, but others are whimsical in nature. You can find shareware fonts in shareware catalogs, bulletin boards, and user groups.

## FREEWARE

Freeware, or public domain fonts, cost nothing. Although many of these fonts lack quality, they are often fun to have because they are unusual in design. You can find freeware on bulletin boards or from user groups.

## ON-LINE SERVICES

If you have a modem, you can obtain fonts from on-line bulletin boards or services such as CompuServe. The fonts you receive from these services can vary in quality. Some fonts may be commercial quality, but others may be freeware or shareware. Some fonts may even be pirated.

Sysops on CompuServe's DTPForum try to ensure that all fonts available in the forum are legal. Most fonts in the DTPForum were uploaded by the creators of the fonts. This forum is a good source for free- and shareware fonts, as well as a good source for desktop publishing discussions and information.

The following on-line services may offer forums for fonts or desktop publishing:

- ★ CompuServe
- ★ GEnie

- ★ UUNET
- ★ Local user-group bulletin boards

**The Coach Says...**

Note that pirated fonts are illegal. Do not download fonts that you know have been stolen. By using pirated fonts, you knowingly violate the law and keep the cost of fonts high. You also risk placing a virus on your system.

# INSTANT REPLAY

In this chapter, you learned to do the following:

INSTANT REPLAY

- ✔ Recognize and use your basic font set
- ✔ Develop a plan for purchasing fonts
- ✔ Determine your font quotient
- ✔ Purchase fonts from a variety of sources

# PART III

# WORKING WITH FONTS

# BEFORE YOU BEGIN

You can have a computer with 100 installed fonts, but that does not make you a professional typographer. You must know how to use those fonts effectively before others will view you as an effective designer. Acquiring the fonts is easy; learning to use them in a professional manner is more difficult.

This chapter teaches you how to avoid the common mistakes that beginning designers are likely to make. By following a few simple rules, you can produce top-quality documents. Specifically, you learn the following:

## GAME PLAN

- Breaking typewriter habits
- Inserting special characters in documents
- Curbing your desire to go wild
- Avoiding overcapitalization and underlining
- Avoiding bad word breaks

## DEVELOPING A FASHION SENSE

The fashion industry dictates what's in and what's out each year. One year short skirts and wide ties may be what's hot. The next year designers may tout longer skirts and narrow ties. These fashion edicts are trends or fads. The fashion industry, however, does follow a general set of fashion guidelines.

You probably are familiar with the following fashion rules:

★ Don't wear white shoes after Labor Day.

★ Never wear brown shoes with a black suit.

★ Under no circumstance ever wear a lime green leisure suit.

Just like the fashion designers, typographers have a basic set of guidelines that they follow. Learning these simple fundamentals will keep you from making fashion faux pas in your documents. Trends in typography, such as the use of specific typefaces, come and go. The basic fundamentals stay in vogue year after year, just as the basic black dress never goes out of style.

## BREAKING TYPEWRITER HABITS

Although you probably have used a wordprocessing program on your computer for some time, you may not realize that typing on a computer is very different than typing on a typewriter. The most obvious difference, of course, is that you don't need to smack the side of your computer screen to advance to the next line.

Seriously, you do need to consider the ways that your computer differs from a typewriter. Your options were limited when you typed on the traditional electric typewriter. The computer, however, gives you more options, and therefore provides you with

more opportunities to make mistakes. By learning the following simple rules, you can create quality documents.

## SPACING BETWEEN SENTENCES

If you took a typing class before the advent of personal computers, you learned to press the spacebar twice after the period at the end of a sentence. Get rid of this rule. You use only one space between sentences on a computer.

Typewriters use monospace fonts. Because each character takes up the same amount of space, an extra space must be inserted after a period so that one sentence can be separated from another. Now that you are using proportional fonts, however, you no longer need to use two spaces. In fact, using two spaces can cause unsightly problems.

The most common problem is known as rivers. *Rivers* are white gaps in your text that visually disrupt the flow of the text. Placing two spaces between sentences also can cause a gap at the end of justified lines of text. Figure 7.1 illustrates why you should use only one space after periods.

Get in the habit of pressing the spacebar only once after all punctuation marks.  If you press the spacebar twice, you may cause unsightly rivers in your text.  If your word processing software has a search-and-replace operation, search for two spaces and replace them with one space.

**Figure 7.1:** Using two spaces can disrupt the flow of text.

**The Coach Says...**

Get in the habit of pressing the spacebar only once after all punctuation. If your word-processing software has a search-and-replace operation, search for two spaces and replace them with one space.

## PLACING QUOTATION MARKS

Always place punctuation marks in the correct place. Nothing says "amateur" as much as misplaced punctuation marks. Don't panic; this is not a grammatical rule. Placement pertains to physical location of the mark, not its grammatical usage.

Quotation marks seem to be the culprit in most punctuation errors. Remember the following rules to create professional-looking documents:

★ Place commas and periods within quotation marks.

★ Place semicolons and colons outside the quotation marks.

★ If more than one paragraph is quoted, place a double quotation mark at the beginning of each paragraph, but only place an ending quotation mark after the last paragraph.

★ Use curly quotation marks, rather than the typewriter ditto marks.

**The Coach Says...**

*Curly quotes* are the marks that you use to specify opening and closing quotation marks. Straight quotation marks are not appropriate to use

in professional documents. In fact, don't even use the straight quotation marks for inch and feet marks; use the single prime and double prime marks. The next section tells you how to locate these special characters.

## USING SPECIAL CHARACTERS

By using some of the special characters in your fonts, you can improve the appearance of your documents. The most commonly used special characters are as follows:

★ Bullets

★ Em and en dashes

★ Ellipses

★ Fractions

★ Superscripts and subscripts

★ Curly quotation marks

### BULLETS

Bullets are useful when you want to show a list of some kind. The preceding list makes use of bullets. When you typed on a typewriter, you probably used asterisks to mark the points on a list. Now, however, you have the option of choosing from an array of bullets. Figure 7.2 shows some of the characters that you can choose to use as bullets.

**Figure 7.2:** Choosing from a variety of bullets.

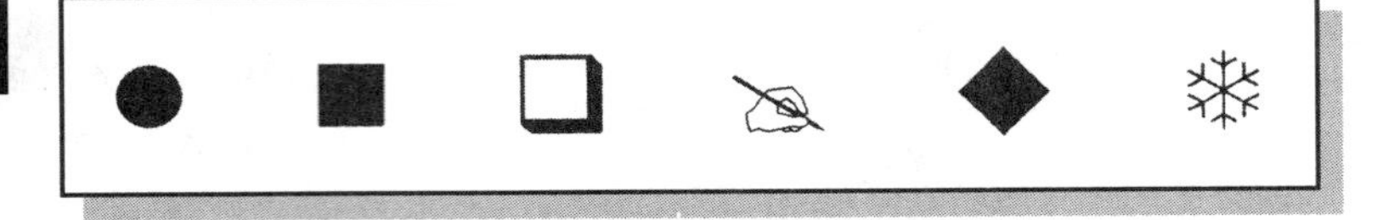

## DASHES

When you type on a typewriter, you can use only the hyphen key to construct dashes. You had to place two hyphens together to form a dash (—). With most software programs, you now have a choice of three types of dashes: hyphens, en dashes, and em dashes.

You use a hyphen to separate words or line breaks. You use en dashes to indicate a range or duration. The following examples illustrate the usage of en dashes:

June 8–January 1

All adults between the ages of 28–40

Work hours are 8:00–5:00

An em dash is twice as long as an en dash—it's roughly the size of an uppercase M. You use em dashes, as in the preceding sentence, when a comma just won't do. Often em dashes replace colons or parentheses.

**The Coach Says...**

If you are using larger than 16-point type, you may find that you should substitute an en dash for em dashes. Often, a full-size em dash just looks too large (see fig. 7.3).

Some desktop publishing programs, such as PageMaker, enable you to set up an em dash with spaces built in on either side. Then you can condense the set width feature to 75 percent. This setting gives the em dash a more reasonable width and provides breathing space on either side of the dash.

Come One—Come All

Come One–Come All

**Figure 7.3:** Use an en dash in larger point sizes.

## ELLIPSES

An ellipsis is a series of three or four periods that indicates that a word or words are missing. You often use elliptical marks when you are quoting from long passages of text. When you type ellipses on a typewriter, you actually type three periods. The ellipses never look quite right when typed on a typewriter. If you put no space between the periods, the ellipsis looks cramped (...). If you place spaces between the periods, the ellipsis looks too airy. The computer, however, treats ellipses as one character. Figure 7.4 illustrates the proper use of ellipses.

… is the time for all good men…

**Figure 7.4:** Using ellipses.

**The Coach Says...**

If the ellipsis character appears to be spaced too tightly, you can type three periods and open up the tracking. This technique often has a more visually appealing spacing than using the ellipsis character.

## FRACTIONS

Typing fractions on a typewriter can be time-consuming, with the result looking rather clunky. You must type the numerator, then the slash (/), and then the denominator. You end up with a fraction that looks like 1/2. You can save time and make your documents look much more professional, however, if you use the built-in fractions that Windows or the Macintosh offer. Later in this chapter you are shown how to use special character sets which include fractions.

**The Coach Says...**

Some font vendors now have special fraction fonts that you can purchase. Adobe has a Helvetica and New Century Schoolbook fraction package, for example. Other fonts have an expert collection available with alternate characters and fractions. Tools, such as Ares' FontMonger, can create fractions from existing Type 1 and TrueType fonts.

## SUPERSCRIPTS AND SUBSCRIPTS

If you have ever tried to type scientific formulas on a typewriter, you should appreciate the superscript and subscript feature of your word processor. Suppose, for example, that you want to type

the scientific name for water. Typing this notation on a typewriter requires that you type an H, then physically move the carriage and type a 2, and then type an O.

Using a word processor, you can simply format the subscript or superscript characters as you type (see fig. 7.5).

[1]Robert Tidrow, "Computers and Man," *Computer Monthly* 113 (1993).

$H_2O$

**Figure 7.5:**

Examples of superscript and subscript characters.

## OTHER SPECIAL CHARACTERS

By using fonts, you have a multitude of special characters at your fingertips. No matter if you are a PC user or a Macintosh user, both platforms enable you to place accent marks over letters in your text. You also can use register marks, trademarks, copyright marks, and special symbols such as check marks and arrows.

The typical keyboard cannot possibly contain graphic representations of all the special characters. The next section shows you how you can find the special characters that you want to use.

# FINDING THE SPECIAL CHARACTER

No matter whether you are using a Macintosh or a PC that is running Windows or DOS, you can find the special character that you want to use by looking on a character set chart. These charts provide you with key combinations that you type to enter the special character from the keyboard.

Windows fonts are based on the ANSI extended character set. *ANSI* is an acronym for the American National Standards Institute. Each character has a corresponding number associated with it. Standard characters that are represented on your keyboard are assigned the numbers between 32 and 126. Special characters that do not appear on the keyboard are assigned the numbers 127 through 255. The DOS and Macintosh character sets are based on the ASCII extended character set.

You can use two methods to insert special characters into your documents. In windows you can enter these characters by pressing and holding down the Alt key and 0, then the unique number code. If you are using a Macintosh, you press a combination of keys. You also can use the character maps in Windows or Key Caps on the Macintosh to see the special characters that are available in a particular font.

## THE KEYBOARD METHOD

Inserting special characters by using the keyboard method is quick and easy. You just need to consult a character chart so that you know which keys to press. Tables 7.1 and 7.2 show the key combinations that you type to use special characters for Windows, Macintosh, and DOS applications.

### Table 7.1: ANSI Character Set Chart for Windows

Text character set
Symbol character set
Wingdings character set
Character Number → 98 b β ♌

| 32 | | | | 64 | @ | ≅ | 🖎 | 96 | | ‾ | ♊ | 0128 | | | ⓪ | 0160 | | | | 0192 | À | ℵ | 🕙 | 0224 | à | ◊ | ➔ |
|---|---|---|---|---|---|---|---|---|---|---|---|---|---|---|---|---|---|---|---|---|---|---|---|---|---|---|---|
| 33 | ! | ! | ✏ | 65 | A | Α | ✌ | 97 | a | α | ♋ | 0129 | | | ① | 0161 | ¡ | ϒ | ○ | 0193 | Á | ℑ | 🕚 | 0225 | á | 〈 | 🡑 |
| 34 | " | ∀ | ✂ | 66 | B | Β | 👌 | 98 | b | β | ♌ | 0130 | ‚ | | ② | 0162 | ¢ | ′ | ⭕ | 0194 | Â | ℜ | 🕛 | 0226 | â | ® | 🡓 |
| 35 | # | # | ✁ | 67 | C | Χ | 👍 | 99 | c | χ | ♍ | 0131 | ƒ | | ③ | 0163 | £ | ≤ | ⭘ | 0195 | Ã | ℘ | ⮰ | 0227 | ã | © | 🡔 |
| 36 | $ | ∃ | 👓 | 68 | D | Δ | 👎 | 100 | d | δ | ♎ | 0132 | „ | | ④ | 0164 | ¤ | ⁄ | ◉ | 0196 | Ä | ⊗ | ⮱ | 0228 | ä | ™ | 🡕 |
| 37 | % | % | 🕭 | 69 | E | Ε | ☜ | 101 | e | ε | ♏ | 0133 | … | | ⑤ | 0165 | ¥ | ∞ | ◎ | 0197 | Å | ⊕ | ⮲ | 0229 | å | ∑ | 🡗 |
| 38 | & | & | 🕮 | 70 | F | Φ | ☞ | 102 | f | ϕ | ♐ | 0134 | † | | ⑥ | 0166 | ¦ | ƒ | ❍ | 0198 | Æ | ∅ | ⮳ | 0230 | æ | ⎛ | 🡖 |
| 39 | ' | ∋ | 🕯 | 71 | G | Γ | ☝ | 103 | g | γ | ♑ | 0135 | ‡ | | ⑦ | 0167 | § | ♣ | ▪ | 0199 | Ç | ∩ | ⮴ | 0231 | ç | ⎜ | 🡐 |
| 40 | ( | ( | 🕿 | 72 | H | Η | 👇 | 104 | h | η | ♒ | 0136 | ˆ | | ⑧ | 0168 | ¨ | ♦ | ◻ | 0200 | È | ∪ | ⮵ | 0232 | è | ⎝ | 🡒 |
| 41 | ) | ) | ✆ | 73 | I | Ι | ✋ | 105 | i | ι | ♓ | 0137 | ‰ | | ⑨ | 0169 | © | ♥ | ⯅ | 0201 | É | ⊃ | ⮶ | 0233 | é | ⎡ | 🡑 |
| 42 | * | ∗ | 🖂 | 74 | J | ϑ | ☺ | 106 | j | φ | 🙰 | 0138 | Š | | ⓿ | 0170 | ª | ♠ | ✦ | 0202 | Ê | ⊇ | ⮷ | 0234 | ê | ⎢ | 🡓 |
| 43 | + | + | 🖃 | 75 | K | Κ | 😐 | 107 | k | κ | 🙵 | 0139 | ‹ | | ❶ | 0171 | « | ↔ | ★ | 0203 | Ë | ⊄ | 🙠 | 0235 | ë | ⎣ | 🡔 |
| 44 | , | , | 📪 | 76 | L | Λ | ☹ | 108 | l | λ | ● | 0140 | Œ | | ❷ | 0172 | ¬ | ← | ✶ | 0204 | Ì | ⊂ | 🙡 | 0236 | ì | ⎧ | 🡕 |
| 45 | - | − | 📫 | 77 | M | Μ | 💣 | 109 | m | μ | ❍ | 0141 | | | ❸ | 0173 | - | ↑ | ✹ | 0205 | Í | ⊆ | 🙐 | 0237 | í | ⎨ | 🡗 |

## ANSI Character Set Chart for Windows

| | | | | | | | | | | | | | | | | | | | | | | | |
|---|---|---|---|---|---|---|---|---|---|---|---|---|---|---|---|---|---|---|---|---|---|---|---|
| 46 | . | . | 📬 | 78 | N | N | ☠ | 110 | n | ν | ■ | 0142 | | | ❸ | 0174 | ® | → | ✹ | 0206 | Î | ∈ | 🙔 | 0238 | î | ⎩ | 🡾 |
| 47 | / | / | 📭 | 79 | O | O | 🏱 | 111 | o | ο | □ | 0143 | | | ❹ | 0175 | ¯ | ↓ | ✵ | 0207 | Ï | ∉ | 🙗 | 0239 | ï | ⎪ | ⇦ |
| 48 | 0 | 0 | 📁 | 80 | P | Π | 🏲 | 112 | p | π | 🞐 | 0144 | | | ❺ | 0176 | ° | ° | ⯐ | 0208 | Ð | ∠ | 🙖 | 0240 | ð | | ⇨ |
| 49 | 1 | 1 | 📂 | 81 | Q | Θ | ✈ | 113 | q | θ | ❑ | 0145 | ‘ | | ❻ | 0177 | ± | ± | ⌖ | 0209 | Ñ | ∇ | 🙐 | 0241 | ñ | 〉 | ⇧ |
| 50 | 2 | 2 | 📄 | 82 | R | Ρ | ☼ | 114 | r | ρ | ❒ | 0146 | ’ | | ❼ | 0178 | ² | ″ | ⟡ | 0210 | Ò | ® | 🙑 | 0242 | ò | ∫ | ⇩ |
| 51 | 3 | 3 | 🗏 | 83 | S | Σ | 💧 | 115 | s | σ | ⬧ | 0147 | “ | | ❽ | 0179 | ³ | ≥ | ⌑ | 0211 | Ó | © | 🙒 | 0243 | ó | ⌠ | ⬄ |
| 52 | 4 | 4 | 🗐 | 84 | T | Τ | ❄ | 116 | t | τ | ⧫ | 0148 | ” | | ❾ | 0180 | ´ | × | ⯑ | 0212 | Ô | ™ | 🙓 | 0244 | ô | ⎮ | ⇳ |
| 53 | 5 | 5 | 🗄 | 85 | U | Υ | 🕆 | 117 | u | υ | ◆ | 0149 | • | | ❿ | 0181 | µ | ∝ | ✪ | 0213 | Õ | ∏ | ⌫ | 0245 | õ | ⌡ | ⬁ |
| 54 | 6 | 6 | ⌛ | 86 | V | ς | ✞ | 118 | v | ϖ | ❖ | 0150 | — | | 🙢 | 0182 | ¶ | ∂ | ✰ | 0214 | Ö | √ | ⌦ | 0246 | ö | ⎞ | ⬀ |
| 55 | 7 | 7 | 🖮 | 87 | W | Ω | 🕈 | 119 | w | ω | ⬥ | 0151 | – | | 🙠 | 0183 | · | • | 🕐 | 0215 | × | ⋅ | ⮘ | 0247 | ÷ | ⎟ | ⬃ |
| 56 | 8 | 8 | 🖰 | 88 | X | Ξ | ✠ | 120 | x | ξ | ⌧ | 0152 | ˜ | | 🙡 | 0184 | ¸ | ÷ | 🕑 | 0216 | Ø | ¬ | ⮚ | 0248 | ø | ⎠ | ⬂ |
| 57 | 9 | 9 | 🖲 | 89 | Y | Ψ | ✡ | 121 | y | ψ | ⍓ | 0153 | ™ | | 🙣 | 0185 | ¹ | ≠ | 🕒 | 0217 | Ù | ∧ | ⮙ | 0249 | ù | ⎤ | ▭ |
| 58 | : | : | 🖳 | 90 | Z | Ζ | ☪ | 122 | z | ζ | ⌘ | 0154 | š | | 🙞 | 0186 | º | ≡ | 🕓 | 0218 | Ú | ∨ | ⮛ | 0250 | ú | ⎥ | ▫ |
| 59 | ; | ; | 🖴 | 91 | [ | [ | ☯ | 123 | { | { | ❀ | 0155 | › | | 🙜 | 0187 | » | ≈ | 🕔 | 0219 | Û | ⇔ | ⮈ | 0251 | û | ⎦ | ✗ |
| 60 | < | < | 🖫 | 92 | \ | ∴ | ॐ | 124 | \| | \| | ✿ | 0156 | œ | | 🙝 | 0188 | ¼ | … | 🕕 | 0220 | Ü | ⇐ | ⮊ | 0252 | ü | ⎫ | ✓ |
| 61 | = | = | 🖬 | 93 | ] | ] | ☸ | 125 | } | } | ❝ | 0157 | | | 🙟 | 0189 | ½ | ⏐ | 🕖 | 0221 | Ý | ⇑ | ⮉ | 0253 | ý | ⎬ | ☒ |
| 62 | > | > | ✇ | 94 | ^ | ⊥ | ♈ | 126 | ~ | ~ | ❞ | 0158 | | | · | 0190 | ¾ | ⎯ | 🕗 | 0222 | Þ | ⇒ | ⮋ | 0254 | þ | ⎭ | ☑ |
| 63 | ? | ? | ✍ | 95 | _ | _ | ♉ | 127 | | | ▯ | 0159 | Ÿ | | • | 0191 | ¿ | ↵ | 🕘 | 0223 | ß | ⇓ | 🡨 | 0255 | ÿ | | ⊞ |

To insert special characters if you are using Windows, follow these steps:

1. Move your cursor to where you want the special character to appear.
2. Press the Num Lock key.
3. Hold down Alt and press 0 and the two- or three-digit code for the character you want to insert. Use the numeric keypad.

   Suppose, for example, that you want to insert a round bullet. You press Alt-0149.
4. After you release the Alt key, the special character appears on-screen.

**Table 7.2**
**ASCII Character Set Chart for DOS and Mac**

| Code | Character | Code | Character | Code | Character | Code | Character |
|---|---|---|---|---|---|---|---|
| 0 | (null) | 35 | # | 153 | ö | 189 | ╜ |
| 1 | ☺ | 36 | $ | 154 | Ü | 190 | ╛ |
| 2 | ☻ | 37 | % | 155 | ¢ | 191 | ┐ |
| 3 | ♥ | 38 | & | 156 | £ | 192 | └ |
| 4 | ♦ | 39 | ' | 157 | ¥ | 193 | ┴ |
| 5 | ♣ | 40 | ( | 158 | ₧ | 194 | ┬ |
| 6 | ♠ | 41 | ) | 159 | ƒ | 195 | ├ |
| 7 | • | 42 | * | 160 | á | 196 | ─ |
| 8 | ◘ | 43 | + | 161 | í | 197 | ┼ |
| 9 | (tab) | 44 | , | 162 | ó | 198 | ╞ |
| 10 | (reserved) | 128 | Ç | 163 | ú | 199 | ╟ |
| 11 | (reserved) | 129 | ü | 164 | ñ | 200 | ╚ |
| 12 | (reserved) | 130 | é | 165 | Ñ | 201 | ╔ |
| 13 | (reserved) | 131 | â | 166 | ª | 202 | ╩ |
| 14 | (reserved) | 132 | ä | 167 | º | 203 | ╦ |
| 15 | – | 133 | à | 168 | ¿ | 204 | ╠ |
| 16 | ► | 134 | å | 169 | ⌐ | 205 | ═ |
| 17 | ◄ | 135 | ç | 170 | ¬ | 206 | ╬ |
| 18 | ↕ | 136 | ê | 171 | ½ | 207 | ╧ |
| 19 | ! | 137 | ë | 172 | ¼ | 208 | ╨ |
| 20 | ¶ | 138 | è | 173 | ¡ | 209 | ╤ |
| 21 | § | 139 | ï | 174 | « | 210 | ╥ |
| 22 | ▬ | 140 | î | 175 | » | 211 | ╙ |
| 23 | ↨ | 141 | ì | 176 | ░ | 212 | ╘ |
| 24 | ↑ | 142 | Ä | 178 | ▓ | 213 | ╒ |
| 25 | ↓ | 143 | Å | 179 | █ | 214 | ╓ |
| 26 | → | 144 | É | 180 | │ | 215 | ╫ |
| 27 | ← | 145 | æ | 181 | ┤ | 216 | ╪ |
| 28 | ∟ | 146 | Æ | 182 | ╡ | 217 | ┘ |
| 29 | ↔ | 147 | ô | 183 | ╢ | 218 | ┌ |
| 30 | ▲ | 148 | ö | 184 | ╖ | 219 | █ |
| 31 | (reserved) | 149 | ò | 185 | ╕ | 220 | ▄ |
| 32 | <space> | 150 | û | 186 | ╣ | 221 | ▌ |
| 33 | ! | 151 | ù | 187 | ║ | 222 | ▐ |
| 34 | " | 152 | ÿ | 188 | ╗ | 223 | ▀ |

To insert special characters if you are using DOS, follow these steps:

1. Move your cursor to where you want the special character to appear.
2. Press the Num Lock key.

3. Hold down Alt and press 0 and the two- or three-digit code for the character you want to insert. Use the numeric keypad.

   Suppose, for example, that you want to insert a percentage sign. You press Alt-037.

4. After you release the Alt key, the special character appears on-screen.

To insert special characters if you are using a Macintosh, follow these steps:

1. Move your cursor to where you want the special character to appear.

2. Press the key combination for the character.

   Suppose, for example, that you want to type a register mark. You press Option-R.

**The Coach Says...**

Although you may memorize the special characters that you use often, you may forget the keystrokes for other characters. Make a reference chart of the keystroke combinations and tape it to the side of your computer. When you need to use a special character, you can look up the keystroke combination quickly.

## THE CHARACTER MAP METHOD

Both Windows and the Macintosh enable you to insert special characters into documents by using a desk accessory. In Windows, this accessory is called the Character Map. On the Macintosh, the

accessory is called Key Caps. By using these accessories, you don't have to look up keystroke combinations on the character set charts. You can insert the desired character by copying it into the Clipboard.

To use the Character Map in Windows, follow these steps:

1. Double-click on the Accessories icon in the Program Manager window.
2. From the Accessories window, double-click on the Character Map icon. The Character Map window appears (see fig. 7.6).

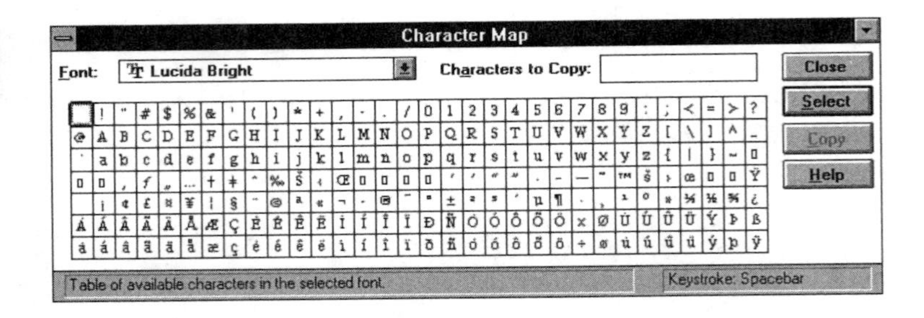


**Figure 7.6:** The Character Map window.

3. Click on the down arrow in the Font list box. Your computer's fonts appear. Note that a TT symbol precedes all TrueType fonts.
4. Select a font. The character set for that font appears.
5. Move the mouse pointer to the character that you want to use, and press and hold down the mouse button. You see an enlarged picture of the character (see fig. 7.7). Note that the keystroke combination you press appears in a box in the lower right of the window.

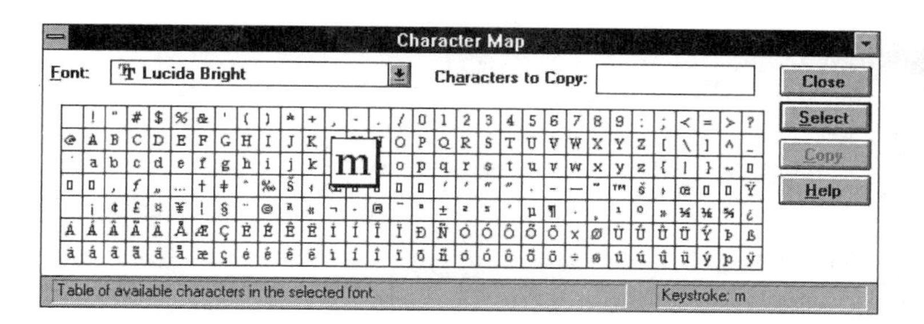


**Figure 7.7:** Looking at the enlarged character.

To insert the character into a document, follow these steps:

1. Double-click on the character you want to insert. You also can click on the character and choose the Select button.
2. After you select the character, click on the Copy button. The character is copied into the Clipboard.
3. Switch to your document, and position the cursor where you want the special character to appear.
4. From the Edit menu, choose Paste. The character appears in your document.

### The Coach Says...

Note that Windows always inserts the special character in 12-point type. If you want the character to appear in another size of type, highlight the character and choose another size.

Note that you may be able to use Windows special characters in DOS applications. Try copying the character to the Clipboard. Windows stores the character in both ANSI and ASCII formats. You then can paste the special character into a DOS application.

## The Coach Says...

This technique does not always work. On some computers, you may find that you can use this technique, but on other computers it just doesn't work. Results are inconsistent. You may find that some characters come across to the DOS application, but do not print. The best thing that you can do is experiment. If this technique does not work, use the ASCII code.

To use the Key Caps accessory on a Macintosh, follow these steps:

1. From your application's Apple menu, select Key Caps. A picture of your keyboard appears on-screen (see fig. 7.8).

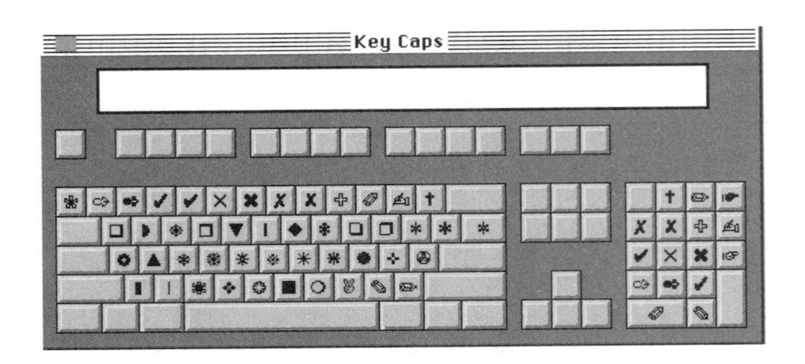


**Figure 7.8:** The Macintosh Key Caps display for the font Zapf Dingbats.

2. From the Key Caps menu, select the font that you want to use. The letters on the on-screen keyboard appear in the font that you choose.
3. You can press a key or click on it with the mouse, and the selected letter appears in the text box at the top of the displayed keyboard.

### The Coach Says...

Note that the Key Caps accessory has four different views. If you press no key, you see the Normal view. If you press Shift (see fig. 7.9), Option, or Shift-Option, however, you see different views of the keyboard.

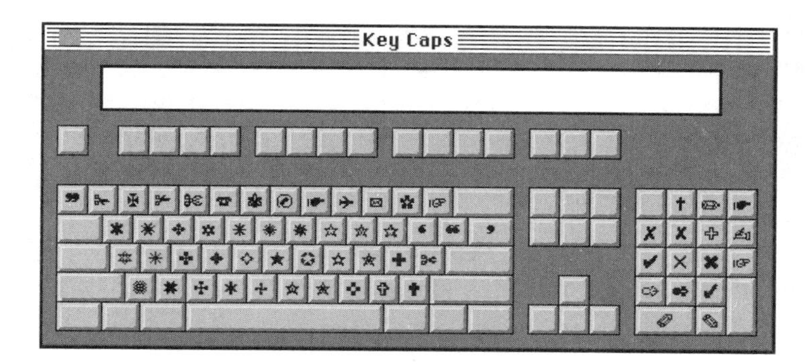


**Figure 7.9:** Viewing Key Caps with the Shift key pressed.

To insert the character into a document, follow these steps:

1. Select the character that you want to use in the sample text box.
2. Copy the character to the Clipboard.
3. Move your cursor to where you want to insert the character and paste the character into the document.

# COMMON TYPOGRAPHICAL ERRORS

As with any new toy that you get, you will be tempted to try out all the options. Don't go overboard, however. Fonts are fun to play with, and you will find a good use for many of them. Just

remember that you don't have to use all the features in one document! You also need to be aware of a few simple rules so that you can create more professional-looking documents. The following section describes the most common pitfalls of the new designer.

## CURBING YOUR DESIRE TO GO WILD

If you haven't had many fonts at your disposal, you will be tempted to overuse them. Don't over-accessorize your document. Use the same restraint that you use when you dress for work. You probably wouldn't wear a gold sequined blouse with a feather boa and black stretch pants—unless of course, you are Madonna. You just wouldn't be taken seriously.

Document design is similar. Do not use more than two different font styles on one document. Using too many typefaces creates an amateurish and cluttered document. Note that each time you change typefaces, your readers must stop and readjust the way they look at your document. Figure 7.10 shows the cluttered look you create when you use too many fonts on one page.

## AVOIDING EXCESSIVE UNDERLINING

Don't overuse underlining. Underlining is useful if you have no other option for calling attention to a word, but it does present problems. Often the line bumps into the descenders and interferes with the readability of the text.

### The Coach Says...

Underlining usually screams "amateur." Underlining was originally used on a typewriter in place of italic. Because the computer offers italic (and bold, and so on), you rarely should use underlining.

**Gigantic Sale!**

NOW THROUGH SATURDAY, FEBRUARY 22

- All merchandise must go! Every item is reduced at least 30 percent.
- Extended shopping hours (8 a.m.-10 p.m.).
- Friendly salespeople to help you find what you are looking for.

*Robinson's Department Store*
*In the Downtown Mall*
Serving the community for 40 years!

**Figure 7.10:** Don't use too many typefaces.

You can use any number of ways to call attention to a word or paragraph. You can use italic, boldface, or a different font. You even can set the text apart from the rest of the document by indenting the margins.

### The Coach Says...

Using underlining is appropriate in some cases. You can use underlining as a design feature, such as a line that appears under subheads. In this case, you use the line feature of your word processor, not the underline style that appears on the menu. The underline style draws a choppy line because it is underlining each separate word. The rule option draws one smooth line (see fig. 7.11).

**Figure 7.11:** The difference between the underline style and rules.

This is underline style.

Use a rule for a cleaner look.

## DON'T BE A BLOCKHEAD

DON'T OVERUSE UPPERCASE LETTERS. Uppercase letters are blockish and hard to read. Studies have shown that readers find it much easier to read passages of text that are printed in upper- and lowercase letters.

All uppercase letters can make your document look as if you are shouting at the readers. By using all uppercase, you may make the reader skip right over the words. People not only read groups of words (phrases), but they also read the shapes of words instead of the actual letters.

People read words by looking at groups of characters, not individual characters. As a person reads a line of type, the eye jumps from word group to word group. This process is called *saccadic movement*. The ascenders and descenders of lowercase letters make it easier for the eye to find focusing points on the line.

### The Coach Says...

Never use all uppercase letters when you are working in a script font. Setting script or any unusual face in all caps makes your type almost impossible to read.

Your word processor may give you the option of using small capitals (see fig. 7.12). Although these capital letters are somewhat easier to read than traditional uppercase letters, you should limit their usage to short lines of type.

UPPERCASE LETTERS ARE HARD TO READ. USE UPPERCASE LETTERS IN SHORT LINES OF TEXT.

SMALL CAPS ARE EASIER TO READ BECAUSE THE EYE CAN EASILY FOLLOW THE BREAKS OF THE LARGER CAPITAL LETTERS.

**Figure 7.12:** Small capital letters.

Note, however, that some word processing and desktop publishing programs just provide a smaller version of the capital letters. These letters tend to be too narrow or thin, and give your document a weird look. Fonts that were designed specifically for small capitals are effective because they were designed to work with normal-sized letters with appropriate width adjustments.

## WIDOWS AND ORPHANS

These terms are not referring to ladies in mourning or children from *David Copperfield.* These terms are actually typographical terms. You can think of the true meaning of these words, however, and relate it to the typographical meaning.

A *widow* is a word or part of a word that stands alone in the last line of a typeset column of text. An *orphan* is a word or words that do not fit at the bottom of a column of text and carry over to the top of the next column. Figure 7.13 illustrates a widow and an orphan.

**Figure 7.13:** Widows and orphans.

A widow is a word or part of
a word that stands alone in the
last line of a typeset column of
*text.*

*text.*
An orphan is a word or single
line that carries to the top of the
next column.

**The Coach Says...**

You can avoid widows and orphans by editing your text. Just by adding or deleting a word or two, you can eliminate the problem.

An orphan is the more serious error to make. A good rule of thumb is to carry over at least two lines of type. Readjust your column break so that you don't leave orphans crying at the tops of your columns.

## BREAKING WORDS

Most word processor programs have built-in hyphenation programs. Don't rely on your computer's hyphenation dictionary, however; you may find it hyphenating in unusual places. Get in the habit of scanning the right side of your text to spot hyphenation problems. The following list shows you what you should look for:

★ Do not hyphenate more than two lines in a row.

★ Do not hyphenate words that contain fewer than six letters.

★ Avoid awkward hyphenations. Examples include susceptibili-ty or a-bove.

★ Do not hyphenate a hyphenated word. An example is co-work-er.

★ Never hyphenate a heading.

You also should be sensitive to line breaks. If you are not justifying your right margin, you probably will come across some bad line endings. You want your right margin to be as even as possible. You may need to adjust the line endings manually.

## The Coach Says...

The subject of hyphenation is a hot topic and many grammar books disagree on the rules. Hyphenation rules are not hard and fast. Proof-reading is the only rule. Make sure that the passage makes sense. Don't rely on the computer's hyphenation dictionary.

Some bad hyphenation examples include the-rapist, tape-stries. These examples can cause the reader to misinter-pret the information. Keep a copy of Strunk & White's *Elements of Style* handy as a reference.

Read your headings to make sure that they break at a logical point. A logical breaking point is at the end of a phrase or name. Bad word breaks may cause a person to misread your heading. Practice Session 7.1 gives you the opportunity to select the appropriate headlines.

### Practice Session 7.1
### Breaking Headings

**Carmel High School Presents Alice in Wonderland**

**A Christmas Tradition: Trimming the Tree**

**The Smith Company Predicts Record Sales**

The second and third headings were examples of good line breaks. The first heading seems awkward because it breaks in the middle of the name of the play. The proper break would have been right after the word "Presents."

## INSTANT REPLAY

This chapter taught you the basic rules of typography. Specifically, you learned the following:

INSTANT REPLAY

- ✔ How many times to press the spacebar
- ✔ Where to place quotation marks
- ✔ Where to find special characters
- ✔ How many typefaces to mix
- ✔ When to use underlining
- ✔ How to avoid all uppercase
- ✔ What widows and orphans are
- ✔ How to avoid bad word breaks

# PUTTING IT ALL TOGETHER

You've learned the rules of the game, you've practiced the fundamentals, and now it's game time. It's time to take the knowledge that you have gained and start producing professional-looking documents.

This chapter not only teaches you step-by-step how to design and set up a document, but also provides tips on how to prepare documents for professional printing.

In this chapter, you learn:

**GAME PLAN**

- ☐ Developing a design plan
- ☐ Using a design checklist
- ☐ Understanding your reader
- ☐ Learning the parts of a page
- ☐ Using copyfitting
- ☐ Setting up a page
- ☐ Understanding kerning

# DEVELOPING A PLAN

Good-looking documents don't just happen. They are the result of good planning. If you give some thought to who your audience is and what you want your message to be, you will have more success than if you have no plan. Spending a little time on the front end of a project can save you time later.

When you sit down before your blank computer screen, you may find it difficult to develop a design plan. After all, you have so many options from which to choose. Before you begin typing your document, however, take a few minutes to consider the document you want to create. By asking yourself a few questions, you can begin to organize your thoughts and ideas.

## QUESTIONS TO ASK

Because good graphic design does not follow an absolute set of rules, you must understand that there is no one right design. Although some fonts and the way in which you arrange text on the page are more appropriate for certain types of documents, you can create several good designs for one project.

How, then, do you develop a plan if you have no set rules to follow? After you have in mind the purpose of your document, you need to ask yourself a few other basic questions. Answering these questions helps you develop a design plan.

### THE PURPOSE

Begin by asking why you need to create the document. What is the purpose? After you define the purpose, you can organize the various bits of information into a logical, cohesive document.

Suppose, for example, that you are developing a brochure. You have before you some text, a few photographs of products, and some pricing information. Assembling this information requires that you view each piece and how that piece fits into the finished picture. Is the purpose of the brochure to introduce new products? Are you advertising special prices? Or is the purpose of the brochure to introduce your company and explain what it produces?

Knowing the answers to these questions helps you prioritize the placement of the various elements in your document. Remember that designing the document is just as much a function of creating a professional quality document as writing the copy that goes in it. Design is as important to the message as the words.

> **The Coach Says...**
> Never view your design as a way to dress up your document. Design must reinforce your message; it shouldn't get in the way. The design must guide the reader from point to point.

## KNOW YOUR READER

If you are creating a report for the board of directors, your approach will be very different than if you are designing a newsletter for the parent/teacher organization at your child's school. Business documents should have a much more serious tone. Keep in mind the level of sophistication of your reader. As you learned in Chapter 1, different fonts evoke different reactions. Choose a font and a page design that complements the reader.

## DECIDE ON FORMAT

After you have decided who your reader is and what the purpose is, you must choose the format to present the information. You can choose from a variety of formats. The following list includes some of the more typical categories of printed presentations:

- ★ Business report
- ★ Brochure
- ★ Newsletter
- ★ Sign or poster
- ★ Invitation
- ★ Advertisement

## SIZING CONSIDERATIONS

You may be limited to a specific size of paper. The overall area that you have to work with greatly influences the design and your choice of fonts. If the area in which you must place your information is small, you should not crowd your text by using overly large and distinctive fonts. An oversized heading will look crowded, and your reader will not want to continue reading.

## REPRODUCING YOUR DESIGN

As you develop your document, consider the way in which you are going to present it to others. Are you going to photocopy the document for office distribution? Are you going to have the document professionally printed?

> **The Coach Says...**
> You can find professional printers by looking in the phone book under "desktop publishing" or "computer services." Professional services use high-resolution printing and output your copy on a Linotronic printer. The Linotronic is a PostScript printing device. You can output your document to film so that you get the highest quality look. Make sure that your printer can handle the file type that you use and that it has the fonts you want to use in your document.

If you plan on having the document printed, consult the printer that you are going to use. The printer can help you set up your document so that the printing process will go smoothly. Sizing and font usage should be discussed.

## A CHECKLIST

Until you consistently ask yourself these questions, you may need a checklist to help you prepare the design of your projects. Figure 8.1 is an example of a worksheet checklist that you can use to help you develop and design successful documents. You can copy this checklist or design one of your own. Then make a photocopy of it and keep it close to your computer.

As you design your documents, ask yourself these questions:

1. What is the purpose of your document?
2. Who is going to read your document?
3. What message are you trying to communicate?
4. Must your document be a specific size?
5. Are you going to have the document professionally printed?

**Figure 8.1:** A design checklist.

# THE PARTS OF A PAGE

A good design utilizes all the elements of a page in the best way possible. Just as you learned the terminology of typography in Chapter 2, this chapter teaches you the terminology you need to understand the elements that make up a page. By understanding each element's purpose, you can better use the space that you have available on the page.

Each page is constructed from basic elements. A page can contain all of the elements or just a few. The way that you use the elements affects the outcome of the design. Think of the page elements as building materials. You are the architect. The materials (page elements) can be used in a number of ways to create many different designs.

## HEADERS AND FOOTERS

The *header* is the information that runs across the top of the page. This information may contain the name of the document, the page number, or the date, or it may be a graphical element, such as a rule or border. A header can contain any information that you want; you are limited only by the size of the page. The information that you place in the header is repeated on the top of every page.

Look at the top of this page. The header for this page includes the name of this book (*The Fonts Coach*). The header also contains a graphical element—a rule. Headers help the readers keep track of where they are.

The *footer* is the information that runs across the bottom of the page. The footer can contain the same types of information and graphics that the header contains. Figure 8.2 illustrates the placement of headers and footers, as well as other parts of the page.

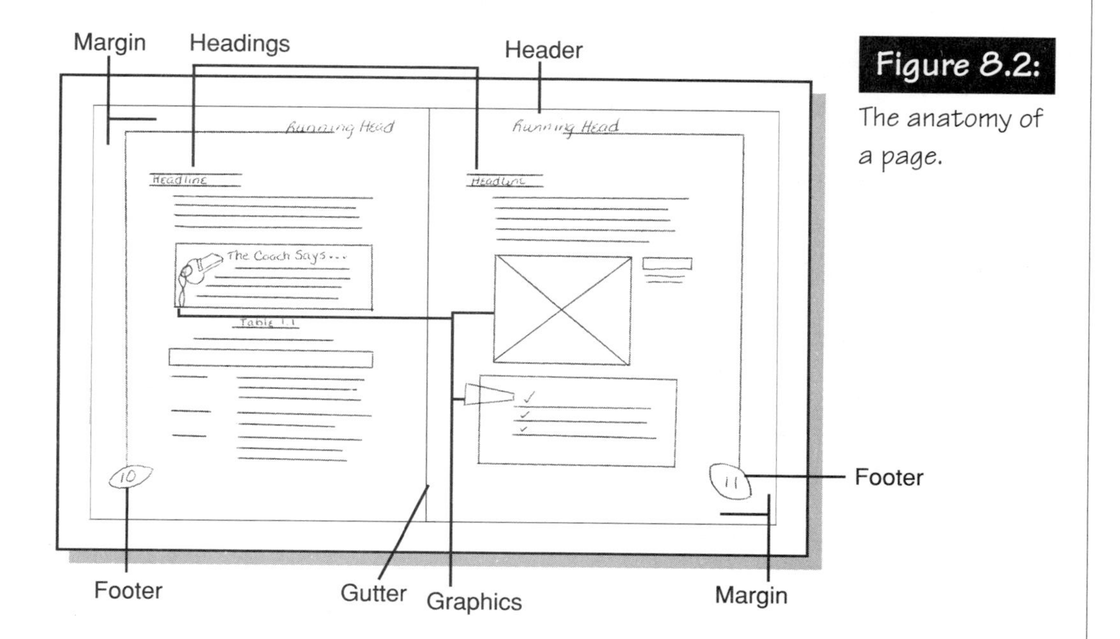


**Figure 8.2:**

The anatomy of a page.

## HEADINGS

The headings, or headlines, call the reader's attention to specific text or tables. As a rule, the longer the text that you create, the better it is to have it broken up with several headings. Headings usually appear in a bold font that may be the same as the text font or in a contrasting font. Suppose, for example, that you are using a serif text font. A good contrasting heading font would be a sans serif bold font, such as Helvetica. Figure 8.2 illustrates headings.

If your text is fairly short, you may need only one level of heading. If your text is long or the subject matter is complicated, you may need to insert other heading levels. This heading level is often called subheadings. Subheadings should be kept short, and should all be of the same tone. Subheadings usually appear on a line by themselves. If you place subheadings on the same line as the text, they are referred to as run-in heads.

## MARGINS

The *margins* of a document are the areas of white space that surround the text. The margins create a focus for the reader's eyes. The margins control how far the text and graphics print from the edge of the printed page. By changing your margin settings, you can change the shape of your printed page.

The inside margin of a page is called the *gutter margin*. The gutter margin is important when you plan to bind the document or place it in a ring binder. You must allow for enough space so that the document can still be read after it is bound.

## GRAPHICS

Graphics are those elements that are not part of the basic body text. A *graphic* can be a picture, a photograph, artwork, clip art, or a table or chart. Graphics can contain text. In fact, the graphic can be totally made up of text.

You often use borders with graphics. A *border* is a decorative frame that encloses the graphic. Borders can be rules, decorative boxes, or shadows. Figure 8.3 shows a shadow-box border.

### The Coach Says...

An excellent companion to *The Fonts Coach* is *The Graphics Coach*. Written by Katherine Murray, *The Graphics Coach* provides step-by-step instructions on when and how to use graphics. The book also discusses the various graphic file formats and how you can convert the formats to use with your specific applications.

**Enrollment Statistics**

| *Class* | *Number of Students* |
|---|---|
| Freshman | 672 |
| Sophomore | 650 |
| Junior | 657 |
| Senior | 600 |

**Figure 8.3:**

A graphic with a drop-shadow border.

# INDENTS

You can use indents to emphasize specific areas of text on the page. The most familiar type of indent is the indent that you see at the beginning of paragraphs. You also can indent text from both the left and right margins. You often see this type of indent when the text is a direct quote from another source or person. A third type of indent is called a *hanging indent*, which is good to use with bulleted or numbered lists. The first line of the hanging indent begins at the page margin, and the following lines are indented. A hanging indent is pictured in figure 8.4.

Types of indents

- ★ First line indent
- ★ Left and right indents
- ★ Hanging indents

**Figure 8.4:**

An example of a hanging indent.

## SIDEBARS

You may have text that pertains to the subject matter, but is not actually part of the flow of the text. You can place this text apart from the general flow of text by placing it in a box or shaded area. Text set apart in such a way is called sidebar text.

> **The Coach Says...**
> The boxed text in this book is sidebar text. Treat sidebar text as a graphic element. Make sure that you don't overuse sidebars. This can be disturbing to readers.

# COPYFITTING

Copyfitting is the technique of writing text to fit the available space. Many people make the mistake of designing a document with no thought to how much text can fit. The result? A document that looks as if it contains too much or too little text.

Writing the text of a document and not giving any thought to available space results in a lot of wasted time. Often you must enlarge or reduce the point size of the fonts to adequately fill the space. You may spend an inordinate amount of time editing and rewriting passages of text.

## THE PROCESS

Unfortunately, the process of copyfitting does require some use of math. The process is not as painful as your high school algebra class, but it does involve some of the same principles.

Copyfitting involves these basic steps:

1. Decide how much space you will specify for each text element.

   Suppose that you are writing a newsletter that will contain three columns of text. Each column is 13 picas wide and seven inches long.

2. Select a point size and font that you want to use. You then determine a character count for that particular font.

3. Type the font's lowercase alphabet and measure the width in points.

   To continue our example, suppose that you select Times New Roman in 10 point size (see fig. 8.5). The font measures 120 points across.

abcdefghijklmnopqrstuvwxyz

**Figure 8.5:** Measuring the width of Times New Roman.

4. Divide this number into 342 to get the number of characters per pica, which is known as character count.

   In this example, divide 342 by 120. The character count for 10 point Times New Roman is 2.9.

5. Multiply the character count by the line length that you want to use.

   In this case, multiply 2.9 by 13. You can fit around 38 characters in 10 point Times Roman on a 13-pica line length.

6. To convert this figure to actual words, divide the number of characters per line by 6. Why six? In the English language, the average word is five characters long. Each word is separated by one space. You add the one space to the average length of five for an average of 6.

   In this example, you divide 38 by 6 for an answer of 6.3.

## Practice Session 8.1

Now practice using your copyfitting skills. Use the techniques that were discussed previously. The correct answers follow.

1. Figure how many characters of 10 point Avant Garde type will fit on a 20-pica line.

   abcdefghijklmnopqrstuvwxyz

2. Figure how many characters of 12 point Bookman type will fit on a 15-pica line.

   abcdefghijklmnopqrstuvwxyz

3. Figure how many characters of 18 point Helvetica will fit on a 30-pica line length.

   abcdefghijklmnopqrstuvwxyz

Answers: 1. 2. 3.

Copyfitting charts also can help you choose the appropriate type size. Table 8.1 shows how many lines of type will fit into how much space.

**Table 8.1: Copyfitting Chart**

| *Column Depth (in inches)* | *Type Size (in points)* 6 | 7 | 8 | 9 | 10 | 11 | 12 | 18 | 20 | 30 |
|---|---|---|---|---|---|---|---|---|---|---|
| 0.25 | 3 | 3 | 2 | 2 | 2 | 2 | 2 | 1 | 1 | 1 |
| 0.50 | 6 | 5 | 5 | 4 | 4 | 3 | 3 | 2 | 2 | 1 |
| 0.75 | 9 | 8 | 7 | 6 | 5 | 5 | 5 | 3 | 2 | 1 |
| 1 | 12 | 10 | 9 | 8 | 7 | 7 | 6 | 4 | 3 | 2 |
| 2 | 24 | 21 | 18 | 16 | 14 | 13 | 12 | 8 | 6 | 5 |
| 3 | 36 | 31 | 27 | 24 | 22 | 20 | 18 | 12 | 9 | 7 |
| 4 | 48 | 41 | 36 | 32 | 29 | 26 | 24 | 16 | 12 | 10 |
| 5 | 60 | 51 | 45 | 40 | 36 | 33 | 30 | 20 | 15 | 12 |
| 6 | 72 | 62 | 54 | 48 | 43 | 39 | 36 | 24 | 18 | 14 |
| 7 | 84 | 72 | 63 | 56 | 50 | 46 | 42 | 28 | 21 | 17 |
| 8 | 96 | 82 | 72 | 64 | 58 | 52 | 48 | 32 | 24 | 19 |
| 9 | 108 | 93 | 81 | 72 | 65 | 59 | 54 | 36 | 27 | 22 |
| 10 | 120 | 103 | 90 | 80 | 72 | 65 | 60 | 40 | 30 | 24 |
| 11 | 132 | 113 | 99 | 88 | 79 | 72 | 66 | 44 | 33 | 26 |

You may want to photocopy the preceding table and keep the chart handy so that you can reference it as you set up documents.

# SETTING UP A PAGE

After you know who your audience is and what your space restrictions are, you can begin to design your document. A good way to start out is to draw a rough sketch of your page.

## DRAWING A THUMBNAIL

A *thumbnail* is a miniature drawing of what your completed page will look like. A thumbnail shows the major design elements of a page. A thumbnail is similar to the map that you scrawl out so that friends can find your house. It shows only the major landmarks.

**The Coach Says...**

Drawing a thumbnail sketch quickly shows you whether your ideas will work. You don't need to be extra neat; this sketch is for your use only. Just use a paper and pencil (see fig. 8.6).

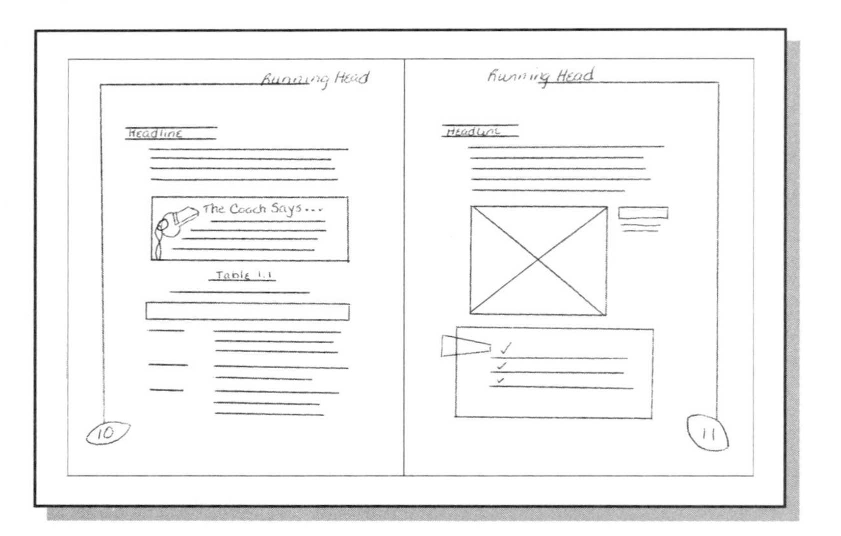


**Figure 8.6:** Thumbnail sketch of a page.

The major landmarks of a page include headlines, text columns, and the white space that surrounds the headlines and text. As you draw the sketch, write in your headlines, and then draw straight lines to represent the text area. Draw several versions. You can experiment with the number of text columns and the placement of headlines or graphic elements, such as pictures.

Avoid bunching visually heavy elements, such as headlines and pictures, in one area of the page. Bunching elements causes the page to look congested in a certain area. The reader also will become bored if forced to read long gray passages of text. Your document looks much more interesting if you give the reader several breaking points.

## PLANNING COLUMNS

Columns contain the text of your document. Before you had the computer and desktop publishing at your disposal, you probably created every document with one column of text. This column of text ran from one margin of your document to the other margin. Now, however, you can create multiple columns of text.

Measure the overall width of your page size to determine the number of text columns you want to use. An 8 1/2-by-11 sheet of paper, for example, contains 612 picas of horizontal space. You can comfortably fit three columns of text in this amount of space.

### The Coach Says...

Remember that you must consider the amount of white space between columns. This white space is called a gutter. Columns that are too close together cause the document to look heavy and make the text difficult to read. By adding additional space between columns, you can open up the text and make it more inviting.

## LINE LENGTH

The number of columns that you set up determines the line length. Remember that the wider the column, the more difficulty the

reader will have in reading the information. The reader's eyes must jump several times across a wide line of text. At the end of the line, the eye must drop down a line and travel across the left side of the column.

Point size also affects the line length that you use. Smaller point sizes can be used in narrow column widths, but using points sizes smaller than 10 point in line lengths over 20 picas is not recommended. Figure 8.7 illustrates the difference in readability between a paragraph of 10 point text and the same text in 12 point. Notice how much easier it is to read the 12 point text.

**Figure 8.7:** Point size helps to determine readability.

Point size also affects the line length that you use. Smaller point sizes can be used in narrow column widths, but using point sizes smaller than 10 point in line lengths over 20 picas is not recommended. This paragraph is 24 picas in length and is set in 10 point Times Roman.

Point size also affects the line length that you use. Smaller point sizes can be used in narrow column widths, but using point sizes smaller than 10 point in line lengths over 20 picas is not recommended. This paragraph is 24 picas in length and is set in 12 point Times Roman.

### The Coach Says...

An easy way to determine whether you are using the correct line length is to type the lowercase alphabet in the point size that you want to use. Line length should be equal to one or one-and-one-half times the length of the alphabet in picas.

## FONT ALIGNMENT

You can create a different look or tone just by changing the alignment of your text. Alignment relates to the way the text aligns on both the left and right margins. Before you used a computer word processing program, you typed text until you came to the right margin. The text aligned on the left margin, but you had no way of aligning the text on the right margin.

With computer word processing programs came the capability of four different alignment settings: flush left, flush right, justified, and centered. Each alignment style is appropriate for a different tone or circumstance.

Ragged right alignment means that the text aligns evenly on the left margin, but the right margin appears uneven (see fig. 8.8). Text that appears ragged right has a more informal, friendly look.

Ragged right alignment means<br>that the text aligns<br>evenly on the left margin,<br>but the right margin<br>appears uneven.

**Figure 8.8:** Ragged right alignment.

Ragged left aligns the text on the right margin, and the left margin appears uneven (see fig. 8.9). Ragged left is a good choice to use for advertisements or for creating special effects. You may want to set ragged left text next to a picture, for example.

**Figure 8.9:**

Ragged left alignment.

Ragged left alignment
means that the text aligns
evenly on the right margin,
but the left margin appears
uneven.

Justified text alignment evenly aligns the left and right margin (see fig. 8.10). Many magazines and newspaper articles appear justified. Justified text is appropriate for books, magazine articles, or newsletters.

**Figure 8.10:**

Justified alignment.

When you use justified text alignment, the left and right margins align so that both sides of the text appear even.

With centered alignment, the text is centered on the specified line length. Centered text is appropriate for headings, pull quotes, or invitations.

### The Coach Says...

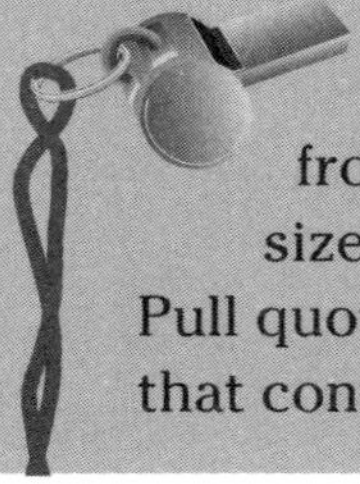

A *pull quote* is a passage of text that you copy from the body of the text and place in a larger point size and style to add visual appeal to the document. Pull quotes are handy tools to use in long documents that contain no other graphical elements.

## FONT SPACING

You can create a beautiful design and choose the appropriate fonts, but your document can still look amateurish if you do not use space properly. As you create a design, you must keep in mind several types of spacing. These spacing considerations are as follows:

- ★ white space
- ★ word space
- ★ letterspace

## WHITE SPACE

White space is the empty space that surrounds the text, pictures, tables, or other graphical elements on a page. You can use white space to make your document more inviting and easier to read. Your text will pack more punch if you use white space creatively.

> **The Coach Says...**
> White space is not necessarily white. If your document is printed on light blue paper, the white space will be blue. The more white space you use, the more of the background color of the paper you see.

So you may be wondering exactly where you should use white space. Remember that readers need breathing space between elements on a page. Try to break up your important elements by inserting white space between them. Table 8.2 provides examples of various types of white space usage.

**Table 8.2:**
**Using White Space**

| *Type of white space* | *Effect* |
|---|---|
| Around a headline | Makes the headline stand out without increasing point size. |
| Between columns | Increases readability of text. The wider the column, the more white space there should be. |
| Indented paragraphs | Useful to break up a text-intense document. |
| Leading between lines | Increases readability. Lines spaced too close together cause a dark looking page. |
| Extra space between paragraphs | Opens up the page. Makes the page more inviting to the reader. |

## WORD SPACE

If you are using justified lines of type in a document, word spacing becomes very important. *Word spacing* is the amount of space that appears between words. If you use too much word spacing, your document will be hard to read and may have unsightly white rivers running through it. Figure 8.11 shows two lines of type. One has very wide word spacing; the other has close word spacing.

This line of text is an example of loose word spacing.

This line of text is an example of close word spacing.

**Figure 8.11:**
Word spacing.

## LETTERSPACE

*Letterspacing* is the space between the individual letters. You can add space or subtract space between letters. In most cases, your automatic letterspacing in your word processing program will be sufficient. Other times, however, you may need to adjust this setting. If you are using all uppercase letters, for example, you will want to move the letters closer together (see fig. 8.12).

SPRING CARNIVAL
SET FOR APRIL 2

SPRING CARNIVAL SET
FOR APRIL 2

**Figure 8.12:**
Letterspacing.

Many times you may not want to set the overall letterspacing for tighter spacing. Sometimes you may only need to move certain letter combinations closer together. Reducing letterspacing in this manner is called *kerning*.

### The Coach Says...

Certain letter combinations require kerning. These letter combinations include Yo, Ya, yo, Y., We, Wo, Wa, we, WA, To, Tr, Ta, Tu, Te, Ty, and TA.

The amount of kerning you must do depends on the type of font you choose. Experiment a little to find the kern setting that looks right for the document you are creating (see fig. 8.13).

**Figure 8.13:**

The effects of kerning.

**WAVE**

**W A V E**

Other times you may want to expand the kerning to achieve a special effect. This effect is very good to use for extending the name of your company or event across the top of a page. Using this type of kerning turn your printed words into a graphical element. You often see this type of effect on business letterhead or advertisements.

# INSTANT REPLAY

In this chapter you learned:

INSTANT REPLAY

- ✔ Developing a design plan
- ✔ Using a design checklist
- ✔ Understanding your reader
- ✔ Learning the parts of a page
- ✔ Using copyfitting
- ✔ Setting up a page
- ✔ Understanding kerning

# MANIPULATING FONTS

This book has given you a wealth of font information, from aesthetic principles through technical jargon. In this final chapter, you get a glimpse of ways you can create a font by using a font editor. You also see how you can manipulate fonts to provide amazing effects with a variety of graphics packages.

This chapter teaches you the following things:

**GAME PLAN**

- ☐ What is a font editor?
- ☐ Types of font editors available
- ☐ How to use a font editor
- ☐ How to test a font editor
- ☐ How to use special effects

# WHAT IS A FONT EDITOR?

If you've ever wondered how fonts are produced, you'll be interested in how a font editor works. The current proliferation of fonts can be directly attributed to the desktop font editors of today's microfoundries. These electronic ateliers create type with cool smarts, not hot metal. As you've learned, the basic font design is only the beginning. It's your responsibility to use well-crafted typefaces to create a beautiful page.

Although desktop publishing page layout programs, such as Aldus PageMaker and Quark Xpress, enable you to condense or expand typefaces, they do not enable you to alter the font outlines themselves. Vector-based drawing packages, however, give you the freedom to freely alter fonts. Programs like Adobe Illustrator and CorelDRAW! provide the capability to make typographical changes—from mild to wild.

The following section introduces you first to font editors.

Can't find a font that fits your needs? Get your hands on a font editor and create one! Do you have little typographical talent, yet still harbor a latent desire to mutilate a classic typeface? If so, you can use a font editor to make changes to an existing font. What is a font editor, exactly? Quite simply, a *font editor* is a program used to create, alter, and fine-tune fonts.

Font editors come in many forms, from inexpensive software (under $100) to big-dollar dedicated systems (such as those from Ikarus). In the following pages, you read about a number of affordable options for the Windows platform.

# WHEN DO YOU USE A FONT EDITOR?

You can use a font editor in many situations. You might want to turn some hand-lettering into a font you can use in a page layout or word processing package, or convert your company's various logos into a font that you can use with any Windows or Macintosh application. Perhaps you want to re-create a typeface you found in an old book you bought at a garage sale.

In these situations, a font editor is exactly what you need to get the job done. While professional type design is a mixture of art, craft, and technology, today's tools enable you to get your feet wet without spending a fortune, making a mess, or blowing up the garage.

You also can use font editors to convert fonts from one format to another. Let's say that you find a particular typeface that is available only in PostScript format, but you use only TrueType typefaces (for whatever reason). A font editor can quickly and easily convert the PostScript font into TrueType form, and vice versa, if you so desire. Two of the most popular font editors with font conversion capabilities are Ares FontMonger and Altsys Fontographer.

## SO YOU WANT TO ROLL YOUR OWN...

Typeface design is not for the faint of heart. It takes serious talent, training, and commitment to design a professional-quality font. You have to start somewhere, though. The first step to creating a font is to have the desire. You've got to want to create. You also need the constitution that will keep you at the workstation until your typeface is ready for prime time.

Sadly, much of what is being done with font editors is merely a regurgitation of previously available typefaces. In the 1990s, cheapie knock-off typefaces are proliferating. On the other hand, the flood of revival typefaces is heartening. These are typefaces that have been lost in specimen books for decades (or centuries) but are suddenly in vogue (or at least now are available).

**The Coach Says...**

What desktop font editors have done, however, is to return the gift of font design to the scribe. In the years before Fontographer and FontStudio, electronic font design was performed on expensive, proprietary workstations. The marketplace was ruled by big foundries that could afford the equipment. Little foundries were shut out. With the advent of Macintosh-based font editors, craftspersons could afford their own tools. This has spurred a flurry of wonderful new typeface designs that look little like anything ever seen before.

To create beautiful and well-engineered fonts, you need to understand more than just how to edit a character outline. You must consider three factors: art, craft, and technology. The art of type design is to create a striking and original font. One that perhaps alludes to, but never plagiarizes, an existing design; unless, of course, you are re-creating a long-lost typeface found in an ancient specimen book.

Re-creating a font isn't an art, however; it's really a craft. It is a craftsperson's work to take an existing design and put it into electronic form. The craftsperson has the responsibility to see that the subtleties and nuances of the artist's design translate through to the printed page. Great care must be taken to ensure that the information stored in the font reproduces properly. This

leads to technology, the final factor. All that art and craft are wasted if the technology is not used correctly. When art, craft, and technology blend together, the fusion often results in a successful face.

The world we live in puts new type designs in our faces at a numbing rate. Never has there been a time when new type has hit the streets as rapidly and been dispersed as widely as today. The successful Neville Brody designs, for example—Industria, Insignia, and Arcadia—that came to market just a short time ago are everywhere.

Take a look at figure 9.1. Do these typefaces look familiar? You've probably run into these faces more than a few times in the past year or so. As you can see, the Neville Brody font collection contains eight typefaces, although they all are variations of those three designs. The A versions contain a handful of alternate characters (such as the lowercase g and t in Industria, as well as the uppercase J, S, and Z in Insignia). New type is everywhere—in print, on packaging, on television, on tee-shirts, on billboards, and riding around on the sides of buses and trucks, and inside passenger trains.

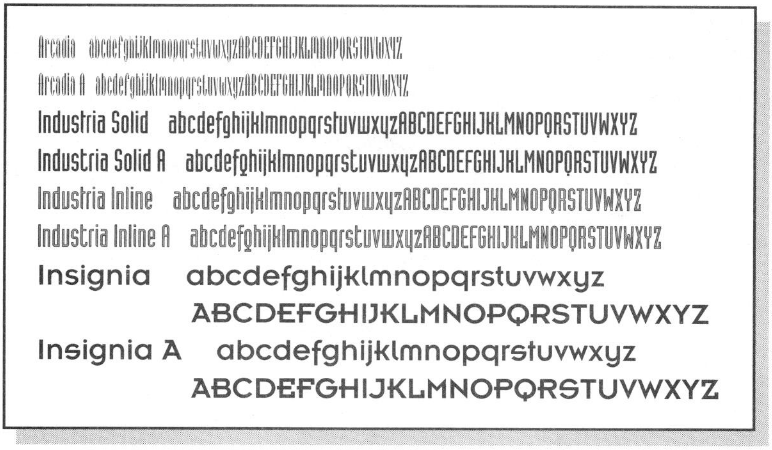

Arcadia abcdefghijklmnopqrstuvwxyzABCDEFGHIJKLMNOPQRSTUVWXYZ
Arcadia A abcdefghijklmnopqrstuvwxyzABCDEFGHIJKLMNOPQRSTUVWXYZ
Industria Solid abcdefghijklmnopqrstuvwxyzABCDEFGHIJKLMNOPQRSTUVWXYZ
Industria Solid A abcdefghijklmnopqrstuvwxyzABCDEFGHIJKLMNOPQRSTUVWXYZ
Industria Inline abcdefghijklmnopqrstuvwxyzABCDEFGHIJKLMNOPQRSTUVWXYZ
Industria Inline A abcdefghijklmnopqrstuvwxyzABCDEFGHIJKLMNOPQRSTUVWXYZ
Insignia abcdefghijklmnopqrstuvwxyz
ABCDEFGHIJKLMNOPQRSTUVWXYZ
Insignia A abcdefghijklmnopqrstuvwxyz
ABCDEFGHIJKLMNOPQRSTUVWXYZ

**Figure 9.1:**
The Neville Brody font collection.

# TYPES OF DESKTOP FONT EDITORS

You can get your typographic designs into electronic form by using an illustration package or a dedicated font editor. Your needs should determine your budget, but in no case should you spend more than a few hundred dollars to get your microfoundry up and running (assuming that you already have a suitable Macintosh or a PC running Microsoft Windows).

The obvious place to start is with an illustration program. You need to have the capability to import a scanned image and to create a vector outline from that image. While an autotracing utility is nice, it merely provides a basis to begin your work. Effective font design and engineering takes lots of handwork and a keen eye. *Vector outlines* are a combination of lines and curves, which are governed by Bézier control points. *Bézier curves* allow font designers to describe font outlines with extreme precision. There's plenty of Bézier tweaking to be done, and the general rule when it comes to the number of Bézier points is: less is more. The cleaner the font outline, the faster it prints and displays and, likewise, the smoother it appears.

**The Coach Says...**

If you get seriously into font design, you soon might end up with a full-blown font editor. Don't think that you have to spend thousands of dollars for software; Fontographer debuted at a street price of less than $300—an incredible bargain for such a powerful program.

## CORELDRAW!

Although not a full-blown font editor per se, CorelDRAW! 3.0 gives users the capability to create both PostScript Type 1 and TrueType fonts. With its comfortable interface, CorelDRAW! is ideal for

creating logotype and for building rudimentary typefaces. The program is limited as a font editor, because it does not provide any control over character kerning, font hinting, and other esoteric and technical issues.

To create a font with CorelDRAW!, you must set up your electronic drawing board specifically for that purpose. You create your character within a space that you create; you set the character's base point by repositioning the rulers. The 0,0 point must be set at the intersection of the character's left side and baseline (see fig. 9.2).

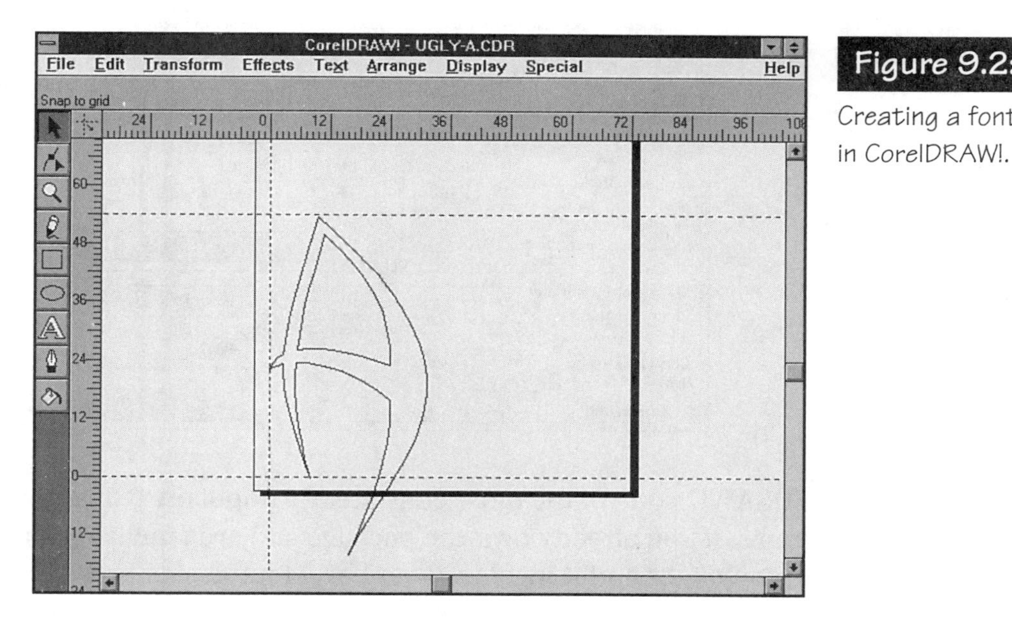


**Figure 9.2:**

Creating a font in CorelDRAW!.

After you have your guidelines down, you then can create a character. Figure 9.2 shows a funky capital "A" with a right stem that tapers off to below the baseline. Each character created must consist of one or a number of objects. The objects cannot overlap. To avoid overlap, you must manually edit the objects.

After you draw the character and run samples to your output device, it's time to export the character to its font. To do this, you must access the Export Font dialog box (for either PostScript

Type 1 or TrueType). Once again, setting things up here is your responsibility (see fig. 9.3). You'll need to name the font, assign the style, set the typeface design size (which varies depending on the size of the original), adjust the grid size (if this is the first character in a new font), and adjust the font's interword spacing. The export box facilitates choosing the character width by providing an Auto Width setting. You must manually choose the character number, although this setting automatically increments on subsequent characters.

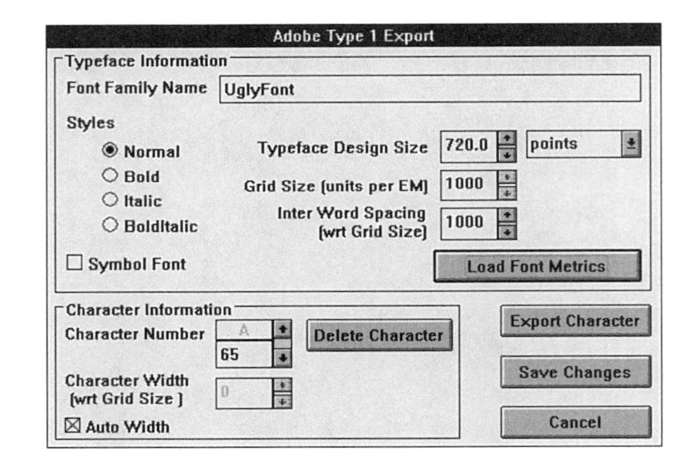


**Figure 9.3:**
Exporting a character to a PostScript font with CorelDRAW!.

CorelDRAW! is one of the most consistently popular PC illustration programs. If you already own the package and are familiar with its tools, you might find it ideal for your early font design projects. After you get more involved, however, you might go looking for more font-editing horsepower.

## ADOBE ILLUSTRATOR

Among vector-based drawing packages, Adobe Illustrator is highly regarded for its technical precision, but it does not offer font creation capabilities like CorelDRAW!. When AI is used in conjunction with a font editor, however, you can achieve better results

than when using CorelDRAW! alone. Because Illustrator files are stored in the native AI format, you do not have to run your files through an export filter (as you do with CorelDRAW!) to be able to import them into a font editor.

The advantages that Adobe Illustrator has over CorelDRAW! show up most favorably in a demanding production environment. Illustrator enables you to have up to 20 different files (or versions of a file) open at one time, while in CorelDRAW! you can have only one file open at a time. Illustrator's precision—to 1/1000 of a point—is legendary and essential to demanding designers.

After you use Illustrator to create the most perfect object possible, you need to use a program such as Ares FontMonger to get that object into a font. In figure 9.4, you see a logo created for the fictional E Corporation. Adobe's Copperplate 33bc was used for all the typography. All type was converted to outlines before saving the file, resulting in six separate objects.

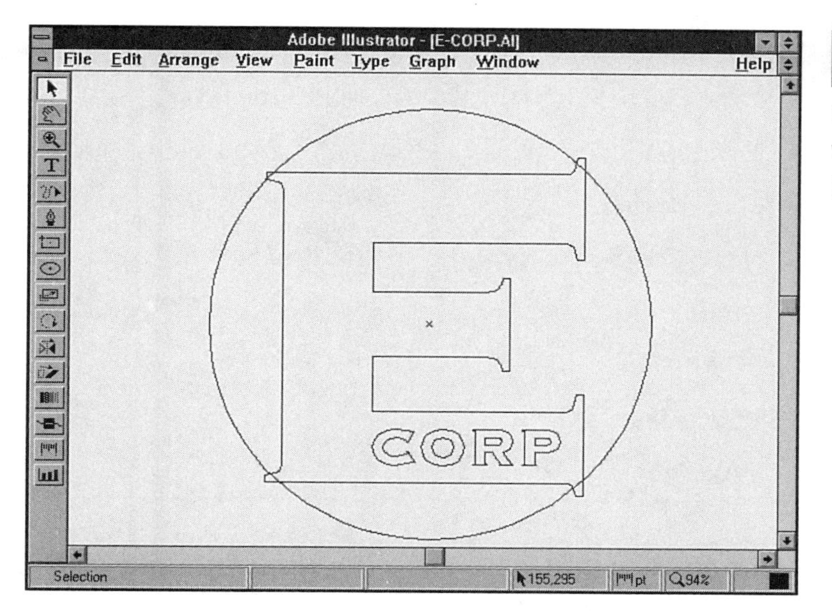


**Figure 9.4:**
Creating a logo with Adobe Illustrator

## ARES FONTMONGER

Ares FontMonger is one of the most useful pieces of software that a Windows desktop publisher can own. The program is designed to convert, enhance, modify, and even create typefaces, and is a perfect complement to Adobe Illustrator or CorelDRAW!. While CorelDRAW! provides the capability to Export TrueType or PostScript Type 1 typefaces, FontMonger's font conversions are far more advanced, and include font hinting, which improves the appearance of laser-printed text at small sizes.

FontMonger can convert fonts from and to a number of formats, as shown in table 9.1.

**Table 9.1:**
**FontMonger Conversion Capabilities**

| *Convert From* | *Convert To* |
|---|---|
| Corel WFN | Mac PostScript Type 1 |
| Intellifonts | Mac PostScript Type 3 |
| LaserMaster LXO | Mac TrueType 1 |
| Mac PostScript Type 1 | NeXT PostScript Type 1 |
| Mac PostScript Type 3 | Nimbus Q |
| NeXT PostScript Type 1 | PC PostScript Type 1 |
| Nimbus Q | PC PostScript Type 3 |
| PC PostScript Type 1 | PC TrueType |
| PC PostScript Type 3 | |
| TrueType | |

Converting fonts between formats is a sticky legal issue. The Ares documentation states that the creation of beautiful and practical typefaces is an art rather than a science, sometimes requiring years of work for the creation of a single typeface. Ares software

supports existing copyrights, the efforts of the typeface designers, and the companies that publish fonts. Check the license agreement for your fonts before using FontMonger to convert or alter them. For more information, contact your fonts supplier directly.

FontMonger presents an affordable choice for those who primarily need to convert fonts between different formats. While the program includes some drawing tools, it is best used in conjunction with an object-oriented drawing package, such as Adobe Illustrator (or CorelDRAW!, although with CD, you must export objects in AI format). FontMonger enables users to import AI format objects and create fonts from those objects. It is best used for altering existing fonts, as well as for creating logo or dingbat fonts. Figure 9.5 shows the E Corp's logo, after being placed into a FontMonger font.

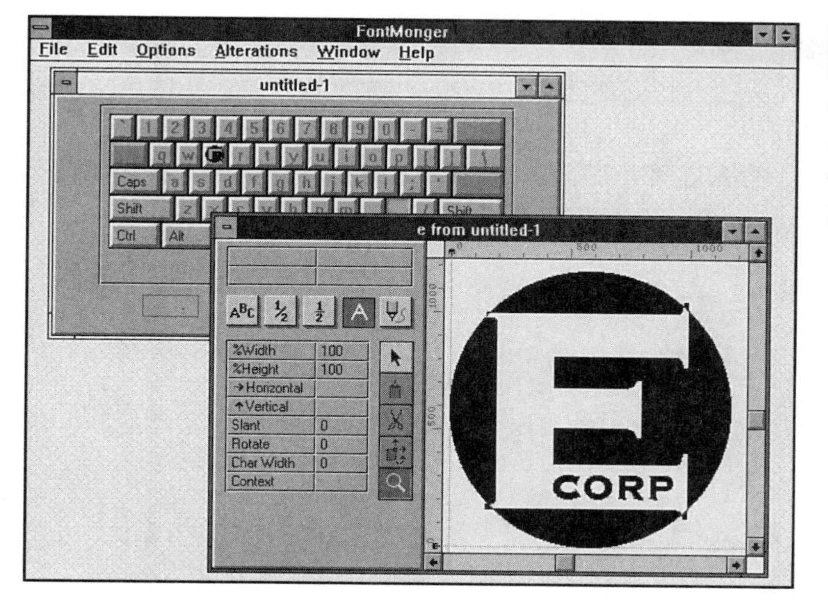


**Figure 9.5:** Creating the E Corporate Logo font with FontMonger.

One of the best features that FontMonger has is the capability to merge multiple object outlines into one. The merge outline feature is a timesaver not offered in either CorelDRAW! or Adobe Illustrator. Overlapping objects are fused together with a minimum of

hassle, and the resulting object remains fully editable (see figs. 9.6 and 9.7).

**Figure 9.6:**

Before merging outlines with FontMonger.

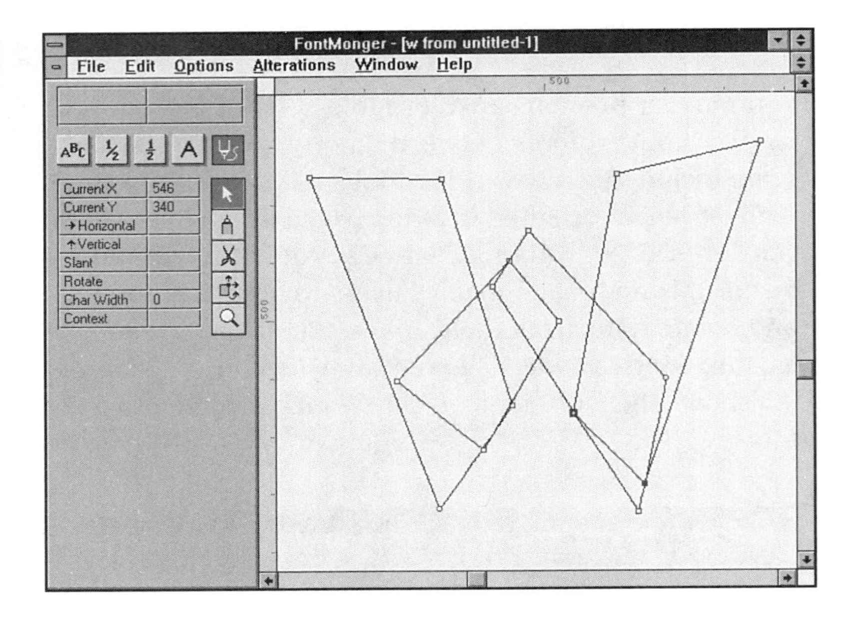


**Figure 9.7:**

After merging outlines with FontMonger.

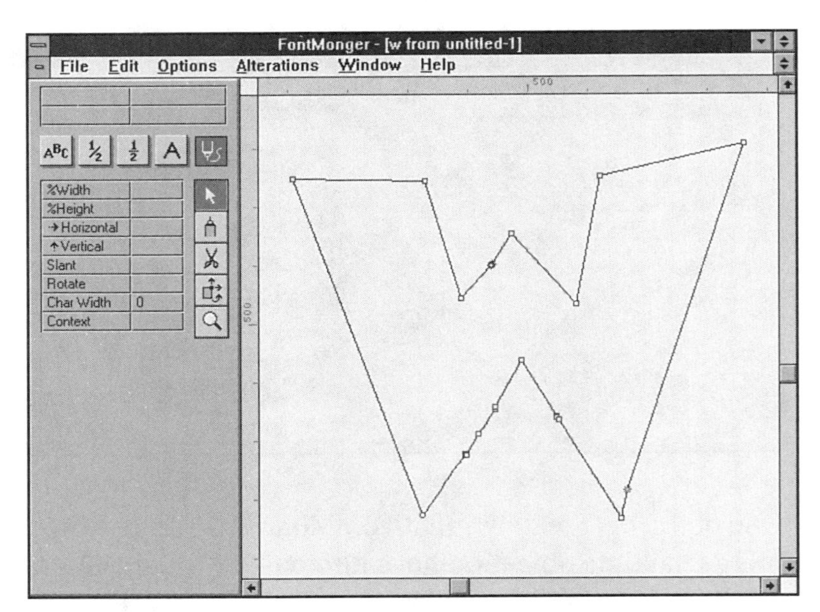

## ALTSYS FONTOGRAPHER

In late 1992, the most popular font editor in the Macintosh world finally made its way to the Windows platform. Altsys Fontographer arrived with all the power and features that made it such a success on the "other side" of the desktop publishing world. Fontographer combines the familiar feel of a vector-based drawing package with the professional tools that serious typophiles crave.

If you are accustomed to working with Adobe Illustrator, Aldus FreeHand, or CorelDRAW!, you will find a similar interface within Fontographer. Characters are defined in an object-oriented manner, rather than in bit-map terms (although the program can create Windows bit-map .FON font files, if desired). While the standard practice is to import bit-map images (of original characters), these are used strictly for tracing. You use Bézier points to control character outlines.

# USING A FONT EDITOR

Putting Fontographer to work is a pleasant task. In the following section, you see how a font is selected, how changes are made, and how the font is saved and put to use. While this is not intended to be a how-to guide, it does give you a taste of how the program works.

When you work with Fontographer, you work with four basic windows. The first of these is the *font window*, where the entire font is displayed in a post office box-like grid. Double-clicking on a character brings up the *character edit window* for that specific character. This is where you make changes to an existing character outline, or create a character outline from scratch. In the *font metrics window*, you can interactively adjust the character kerning pairs. In the *bit map window*, you can wile away your time editing character bit maps (an unnecessary task if you are using PostScript or TrueType fonts).

## SELECTING THE FONT

Fontographer works with fonts in its native .FOG format. If you start with PostScript or TrueType format fonts, it automatically converts them into FOG files. Opening a font is as simple as double-clicking on its filename from within the Open Font dialog box. FOG files open immediately, while PS or TT files take a few minutes to convert. The speed at which the files convert is, of course, related to the speed of your computer (as well as the complexity of the font). Consequently, the faster your computer, the faster the files convert.

After a font file is open, the characters appear in the font window. In figure 9.8, you see the font AlKochAntiqua-Demi, designed by Randall Jones and available from Alphabets, Inc., displayed in all its glory. This font comes with Fontographer 3.5, along with a smattering of other interesting typefaces. The following paragraphs show you how to use the existing numerical characters to create a new fraction character.

**Figure 9.8:**

AlKochAntiqua-Demi in the font window

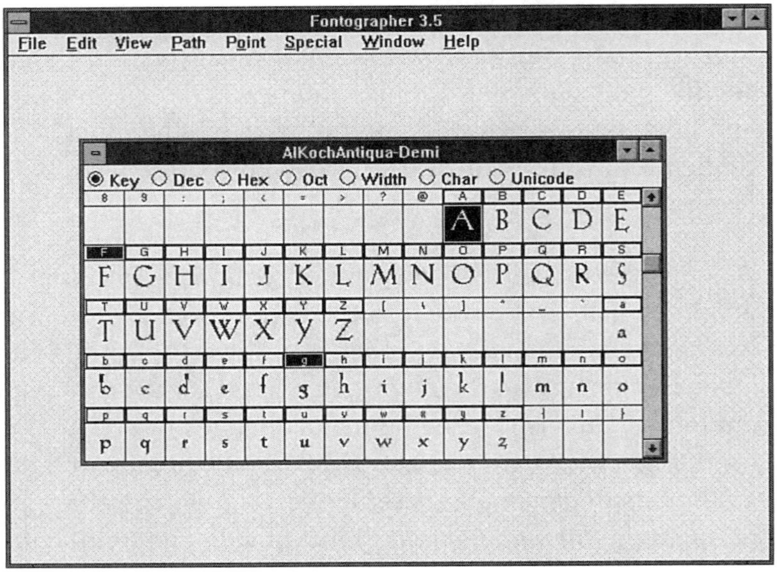

## CREATING A FRACTION

To create a fraction, you need to use three characters: the numerator, the denominator, and the slash. These three characters are copied and pasted from their original windows into the new fraction character—in this case, one-half—accessed via the ALT-0189 key combination. Figure 9.9 displays the characters in their original character edit windows, along with the empty one-half character edit window.

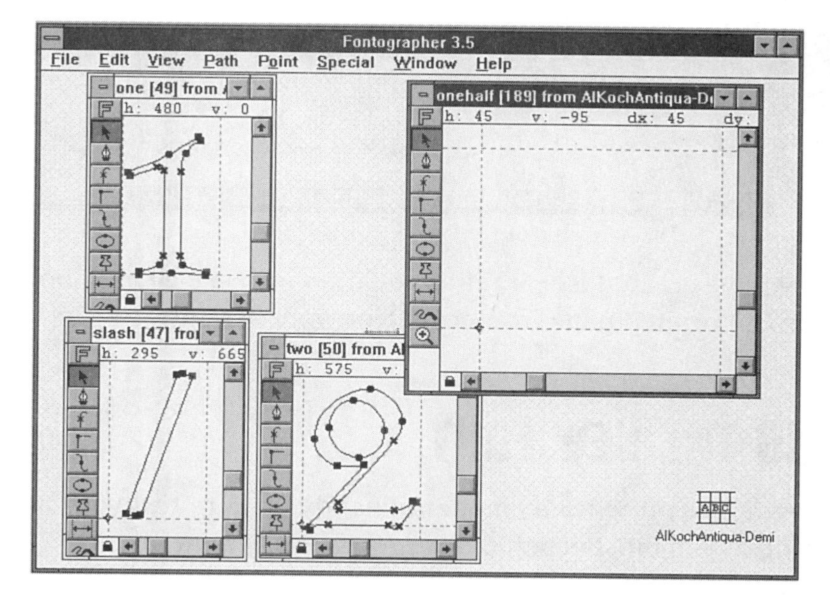


**Figure 9.9:**
Getting ready to create a fraction

You don't have to actually open up a character edit window to copy the character within. You simply can select the character from the font window, copy it, go to the new character edit window, and paste. In figure 9.10, the one, the two, and the slash have been copied and pasted into the one-half character edit window, which has been maximized and zoomed up for more control. Both the one and the two have been proportionally scaled down by 50 percent. The character width has been set at 500. At this point, it's a good idea to save the font so that any changes are not lost.

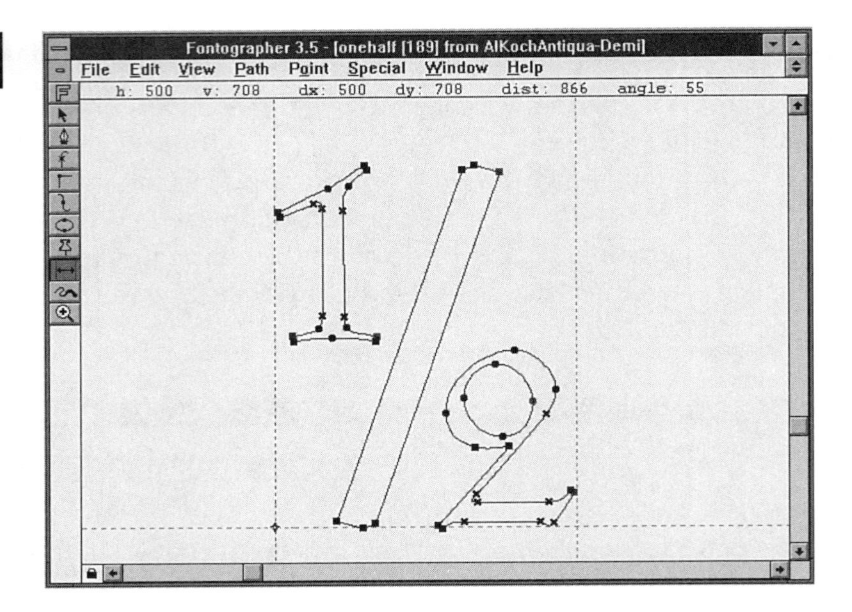


**Figure 9.10:**

A new fraction.

After you've created a character, you'll want to see what it looks like on the printed page (rather than on your screen).

## TESTING THE NEW FONT

Fontographer provides a number of handy ways to test your new characters—in most cases, before you actually generate the PostScript or TrueType fonts. This testing is done through the Print Sample dialog box. A variety of samples are available:

- ★ **Sample Text.** You can either use the ubiquitous default sample, "The quick brown fox jumps over the lazy dog," or come up with your own gibberish.
- ★ **PostScript File.** This option prints a PostScript font file.
- ★ **Key Map.** Just what it sounds like! This option prints out a road map to all the characters in the chosen font, along with their keyboard positions. It also provides width and offset information, along with the ASCII code required for special characters.

★ **Kerning Pairs.** Prints a list of pairs with their kerning values, but not the actual pairs themselves.

★ **Individual Characters.** Issues a close-up look at each chosen character, including precise coordinate information. Also enables you to print samples in a number of sizes.

## FINE-TUNING THE FONT

After you test the font to see that all the necessary characters are there, and are at least legible, you can begin fine-tuning the font. This stage is often overlooked by neophyte typographers—and it shows! While in the process of evaluating the font, pay close attention to kerning pairs. Of course, if the font was created with no kerning whatsoever, the end result might not be acceptable for professional quality work. (Not sure what kerning is all about? Check out Chapter 8.)

With Fontographer, you create and edit kerning pairs from within its Font Metrics Window. In figure 9.11, you can see that the "Wa" character pair is set to kern –129 units. This means that the two characters will be 129/1000s of an EM closer together than their character widths would normally allow. Notice how the W hangs over the a? This would not be as apparent if the guideline were not there! Kerning is a very subtle craft.

### The Coach Says...

PFMEdit is a freeware program that provides access to any PostScript Type 1 font's kerning pairs. With PFMEdit, you can alter existing pairs or create new pairs of your own. You can find it in the CompuServe DTPFORUM libraries.

**Figure 9.11:**

The "Wa" kern pair.

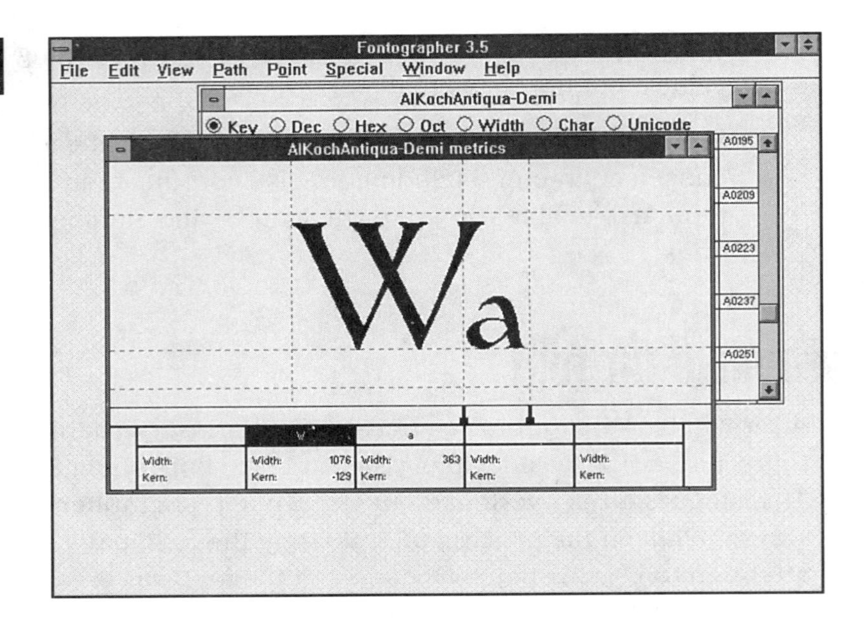


## GENERATING FONTS

With everything in place, the last step—with a twist—is to actually generate the PostScript or TrueType font. Why is there a twist? The creation of a successful typeface has a significant revision cycle. Like a football team that adjusts the lineup and changes its strategy while scrimmaging, you can generate a succession of slightly different files as you go through an extended fine-tuning process. The Generate Fonts dialog box is shown in figure 9.12.

**Figure 9.12:**

Generating a PostScript font.

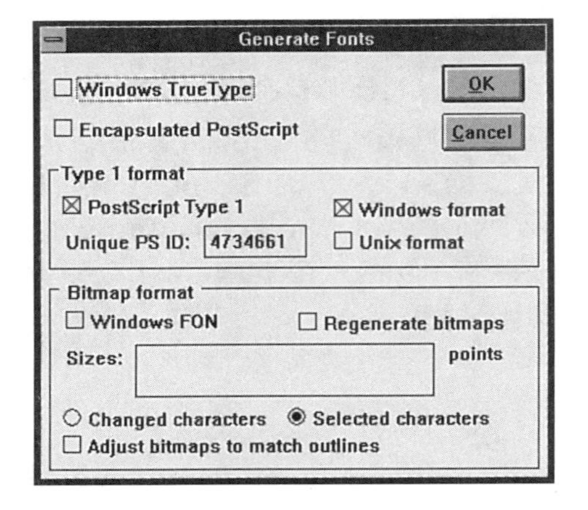

As with most things, Fontographer makes font generation a pleasant, straightforward task, whether you are generating PS, TT, or FON fonts. And once again, the speed at which it creates the font file is dependent on both the speed of your computer and the complexity of the font itself. After the font is created, you load it with the appropriate control panel.

## USING SPECIAL EFFECTS

Creating a special font is one thing, and creating a special type effect is something different altogether. You can use the same illustration programs—Adobe Illustrator, CorelDRAW!, or whatever's handy—to create distinctive type effects. These can range from the pedestrian to the really wild. Your imagination is your most important tool, no matter what program (or programs) you use.

## A LIBRARY OF SPECIAL EFFECTS

You can do practically anything to a piece of type. Of course, it's possible that you will have a tough time actually being able to read what you have created, so tread carefully! You can create many special effects with a variety of programs. Not all effects are possible with all programs, and some effects can take far longer to create than others. This section isn't intended to be a seven-course step-by-step tutorial. It's more a sampler of tasty typographical treats. Bon appetit!

Obviously, different genres of programs have the capability to create different effects. In many cases, you might use more than one program to get the job done. The illustrations in this library of special type effects (see figs. 9.13 through 9.25) make use of a number of programs. Among the programs used are Adobe Illustrator for Windows 4.0, CorelDRAW! 3.0, Crystal 3-D Designer, and Image-In-Color.

**Figure 9.13:**

Flat beveled type.

**Figure 9.14:**

Full beveled type.

**Figure 9.15:**

Chrome type.

**Figure 9.16:**

Diffusion type.

**Figure 9.17:**

Drop shadow variations.

**Figure 9.18:**

Embossed type.

**Figure 9.19:**

Enveloped type.

**Figure 9.20:**

Neon type.

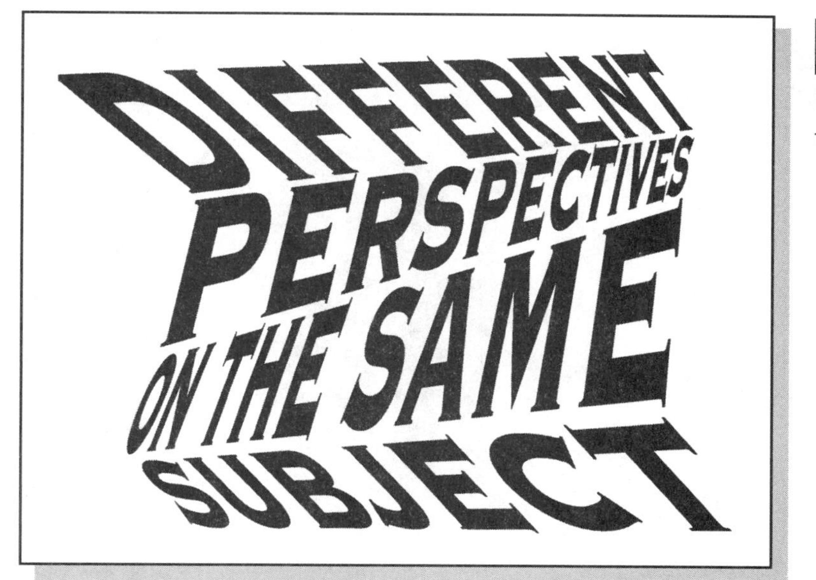


**Figure 9.21:**

Perspective type.

**Figure 9.22:**

Text on a path.

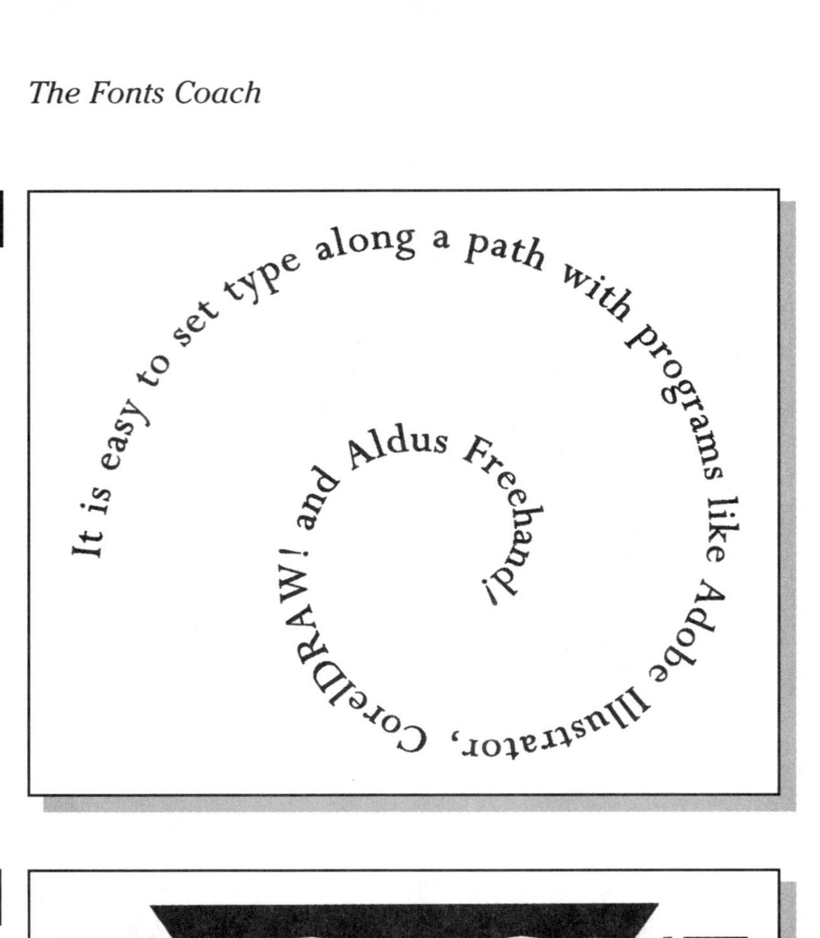


**Figure 9.23:**

Two-tone type.

**Figure 9.24:**

3D text, created with Crystal 3D Designer.

**Figure 9.25:**

Tube type.

# INSTANT REPLAY

In this chapter, you learned about the basics of font editors, and (hopefully) enjoyed our little library of special effects.

INSTANT REPLAY

- ✔ What is a font editor?
- ✔ Types of font editors
- ✔ Using a font editor
- ✔ Testing the new font
- ✔ Using special effects
- ✔ A library of special effects

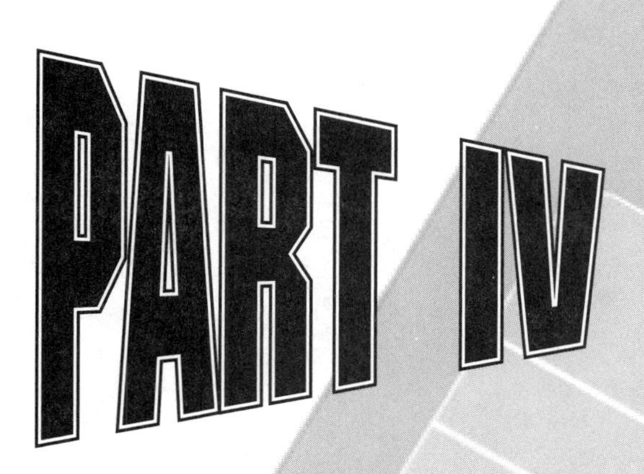


# APPENDIXES

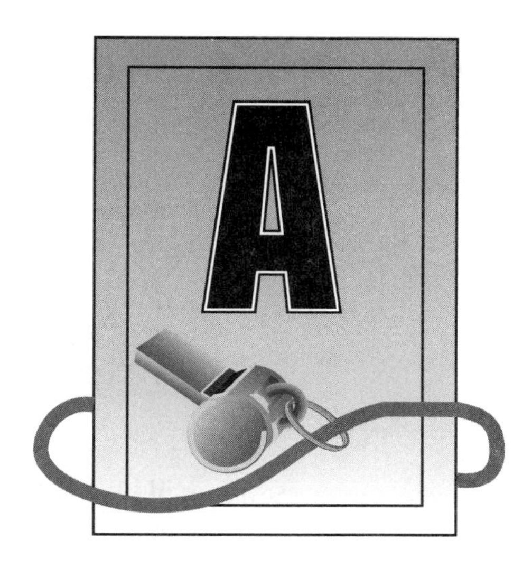

# VENDOR LIST

This appendix contains a list of font manufacturers and companies that have developed font utilities or font managers. Although the editors have tried to maintain accuracy as much as possible, note that phone numbers or addresses can change without notice.

Most of the companies listed provide free literature upon request. You may find it helpful to call these companies to receive the literature before you begin purchasing fonts. You can compare prices and services to make sure that you are getting the most for your money.

**A**

Acute Systems
Box 37
Algonquin, IL 60102

Adisys Data Info. Systems
#200-25 Alexander St.
Vancouver, BC
CANADA V6A 1B2

Adobe Systems, Inc.
1585 Charleston Rd.
P.O. Box 7900
Mountain View, CA 94039-7900
(800) 833-6687
(415) 961-4400

AGFA Compugraphic
90 Industrial Way
Wilmington, MA 01887
(800) 424-8973
(508) 658-5600

Altsys Corp.
269 W. Renner Rd.
Richardson, TX 75080
(214) 680-2060

Architext, Inc.
121 Interpark Blvd., #208
San Antonio, TX 78216-1808
(512) 490-2240

Ares Software Corp.
561 Pilgrim Dr., Suite D
Foster City, CA 94404
(415) 578-9090

Atech Software
5964 La Place Ct., #125
Carlsbad, CA 92008
(619) 438-6883

Autologic, Inc.
1050 Rancho Conejo Blvd.
Thousand Oaks, CA 91320
(805) 498-9611

**B**

BAF Font Foundry
P.O. Box 700
Monsey, NY 10952
(800) 626-2344

Beyond Words
11 Belle Ave.
San Anselmo, CA 94960
(415) 721-2000

BitStream, Inc.
201 First St.
Cambridge, MA 02142
(800) 522-3668
(617) 497-6222

Blue Sky Research
534 S.W. Third Ave.
Portland, OR 97204
(503) 222-9571

Brooks/Cole Publishing Co.
511 Forest Lodge Rd.
Pacific Grove, CA 93950-5098
(408) 373-0728

**C**

Casady & Greene
22734 Portola Dr.
Salinas, CA 93908
(800) 359-4920
(408) 484-9228

Castcraft Software, Inc.
3649 W. Chase Ave.
Skokie, IL 60076
(708) 675-6530

Computer Peripherals, Inc.
667 Rancho Conejo Blvd.
Newbury Park, CA 91320
(800) 854-7600
(805) 499-5751

Custom Applications, Inc.
900 Technology Park Dr.
Billerica, MA 01821
(800) 873-4367

**D**

Data Transforms, Inc.
616 Washington St.
P.O. Box 300458
Denver, CO 80203
(303) 832-1501

Davka Corporation
845 N. Michigan, Suite 843
Chicago, IL 60611
(312) 944-4070

Digi-Fonts, Inc.
528 Commons Dr.
Golden, CO 80401
(800) 242-5665
(303) 526-9435

Digital Typeface Corp.
9965 W. 69th St.
Eden Prairie, MN 55344
(612) 943-8920

**E**

Eicon Technology Corp.
2196 32nd Ave.
Lachine, Quebec
CANADA H8T 3H7
(514) 631-2592

Elesys, Inc.
528 Weddell Dr.
Sunnyvale, CA 94089
(408) 747-0233

Elfring Soft Fonts
P.O. Box 61
Wasco, IL 60183
(708) 377-3520

Elite Business Applications, Inc.
412-3 Headquarters Dr.
P.O. Box 300
Millersville, MD 21108
(800) 942-0018
(301) 987-7200

Elixir Technologies Corp.
P.O. Box 1559
Ojai, CA 93024
(805) 640-8054

EmDash
P.O. Box 8256
Northfield, IL 60093
(708) 441-6699

Emigre Graphics
4475 D St.
Sacramento, CA 98519
(916) 415-4344

ERM Associates
29015 Garden Oaks Ct.
Agoura Hills, CA 91301
(800) 288-3762

Everex Systems, Inc.
48431 Milmont Rd.
Fremont, CA 94538
(800) 821-0806

**F**

Font Bank
2620 Central St.
Evanston, IL 60201
(708) 328-7370

The Font Company
7850 East Evans, Suite 111
Scottsdale, AZ 85260
(602) 998-9711

Font World
2021 Scottsville Rd.
Rochester, NY 14623-2021
(716) 235-6861

**G**

Giampa Textware Corp.
1340 East Pender St.
Vancouver, BC
CANADA V5L 1V8
(604) 253-0815

Glyph Systems, Inc.
P.O. Box 134
Andover, MA 01810
(508) 470-1317

Good Software Corp.
13601 Preston Rd., #500W
Dallas, TX 75240
(800) 925-5700

Graphitec
9101 W. 123rd St.
Palos Park, IL 60464
(708) 448-6660

**H/I**

Hewlett-Packard Co.
19310 Pruneridge Ave.
Cupertino, CA 95014
(800) 752-0900

IBM Corp.
6300 Diagonal Hwy.
Boulder, CO
(303) 924-7838

Image Club Graphics
1902 11th St., Suite 5
Calgary, Alberta
CANADA T2G 362
(800) 661-9410

Innovage International Corp.
21436 N.E. Union Hill Rd.
Redmond, WA 98053
(206) 462-5787

IQ Engineering
685 N. Pastoria Ave.
Sunnyvale, CA 94086
(800) 765-3668
(408) 733-1161

**K/L**

K-Talk Communications, Inc.
30 West First Ave., 100
Columbus, OH 43201
(614) 294-3535
(614) 294-3704

LaserGo, Inc.
9369 Carroll Park Dr., Suite A
San Diego, CA 92121
(619) 450-4600

LaserTools Corp.
1250 45th St., #100
Emeryville, CA 94608
(800) 767-8004
(510) 420-8777

Letraset G.D.S.
40 Eisenhower Dr.
Paramus, NJ 07653
(201) 845-6100

Linotype-Hell
425 Oser Ave.
Hauppauge, NY 11767
(800) 633-1900
(516) 434-2000

Lotus Selects
P.O. Box 9172
Cambridge, MA 02139
(800) 635-6887

LTI Softfonts International, Inc.
14742 Beach Blvd., Suite 440
La Mirada, CA 90638
(714) 739-1453

**M/N**

MacTography
326D N. Stonestreet Ave.
Rockville, MD 20850
(301) 424-1357

Mephistopheles Systems Designs
3629 Lankershim Blvd.
Hollywood, CA 90068-1217
(818) 762-8150

Metro Software
1870 W. Prince Rd.
Tucson, AZ 85705
(800) 621-1137

MicroGraphix
1303 Arapahoe
Richardson, TX 75081
(800) 733-3729
(214) 234-2694

MicroLogic Software, Inc.
1351 Ocean Ave.
Emeryville, CA 94608
(800) 888-9078
(510) 652-5464

Microsoft Corp.
One Microsoft Way
Redmond, WA 98052-6399
(206) 882-8080

MonoType Typography
53 W. Jackson Blvd., #504
Chicago, IL 60604
(800) 666-6897
(312) 855-1440

NEC Technologies, Inc.
1414 Massachusetts Ave.
Boxborough, MA 01719
(508) 264-8000

## O/P

Olduvai Corp.
7520 Red Road, Suite A
South Miami, FL 33143
(305) 665-4665

Pacific Data Products
9125 Rehco Rd.
San Diego, CA 92121
(619) 552-0880

Page Studio Graphics
3175 N. Price Rd., Suite 1050
Chandler, AZ 85224
(602) 839-2763

Personal Computer Products, Inc.
10865 Rancho Bernardo Rd.
San Diego, CA 92127
(800) 262-0522 (CA)
(800) 225-4098 (outside CA)

Personal TeX, Inc.
12 Madrona Ave.
Mill Valley, CA 94941
(415) 388-8853

PMWare, Inc.
346 State Pl.
Escondido, CA 92029
(800) 845-4843
(619) 738-6633

Power Up Software
2929 Campus Dr., #400
San Mateo, CA 94403
(800) 851-2917
(415) 345-0551

**Q/S**

QMS, Inc.
One Magnum Pass
Mobile, AL 36618
(205) 633-4300

QualiType
29209 North Western Hwy., #611
Southfield, MI 40834
(313) 822-2921

Silver Graphics
P.O. Box 485
Pennington Gap, VA 24277
(703) 546-3800

SoftCraft, Inc.
16 N. Carroll St., #500
Madison, WI 53703
(800) 351-0500

Sonnet Technologies, Inc.
18004 Sky Park Circle, Suite 260
Irvine, CA 92714
(714) 261-2800

Studio 231
231 Bedford Ave.
Bellmore, NY 11710
(516) 785-4422

SWFTE International, Ltd.
P.O. Box 219
Rocklind, DE 19732
(302) 234-1740

**T/U/V**

Treacyfaces, Inc.
P.O. 26036
West Haven, CT 06516
(203) 389-7037

TypeWorks, Inc.
820 Cummings St.
Abingdon, VA 24210
(800) 221-9920

TypeXpress
1510 Fenol Lane
Hillside, IL 60102
(800) 343-4424

UDP
1309 Laurel Ave.
Manhattan Beach, CA 90266
(213) 545-5767

URW
#4 Manchester St.
Nashua, NH 03060
(603) 882-7445

Varityper
11 Mt. Pleasant Ave.
East Hanover, NJ 07936
(800) 631-8134
(201) 887-8000

VN Labs
4320 Campus Dr., Suite 114
Newport Beach, CA 92660
(714) 474-6968

VS Software
209 West Second
Little Rock, AR 72216
(501) 376-2083

**W/X/Z**

Weaver Graphics
5161 S. Highway A1A
Melbourne Beach, FL 32951
(407) 728-4000

Xerox Corp.
Xerox Square
Rochester, NY 14644
(716) 423-5078

Zebra Technologies, Inc.
300 Corporate Woods Pkwy.
Vernon Hills, IL 60061-3109
(708) 634-6700

Zenographics
4 Executive Circle
Irvine, CA 92714
(714) 851-6352

ZSoft Corp.
450 Franklin Rd., Suite 100
Marietta, GA 30067
(404) 428-0008

# GLOSSARY

Typesetting today is a merging of the centuries-old art of typography and the latest computer technological breakthroughs. This glossary contains terms from both areas. Keep this appendix handy as a desktop reference to look up definitions of terms that you don't understand.

**alignment.** The arrangement of the printed lines of type on a page. Alignment can be flush left, flush right, or justified.

**ascender.** The part of a letter that extends above the body of the letter, such as the upward stroke of a d or b.

**ASCII.** An acronym for American Standard Code for Information Interchange. This code uses a numbering scheme to denote characters and control commands. ASCII codes 32 through 255 are used to print characters.

**backslant face.** A typeface that slants backward. This face is opposite of italic.

**bad break.** An incorrect line ending, such as bad hyphenation or an orphan or widow. See also *orphan* and *widow*.

**baseline.** The imaginary line on which the bottoms of characters sit. The descending strokes of characters fall below the baseline.

**BBS.** See *bulletin board system.*

**bit map.** A grid of dots that make up the image of a character. Each character is made up of a unique pattern of dots.

**bleed.** An area of a design that goes beyond (bleeds off) the edge of a page. Bleeds most often involve graphics or colors.

**body copy.** Text that makes up the body of a document. Body copy is usually between 8 and 12 points in size.

**border.** A decorative line or design that surrounds printed material or photographs.

**bullet.** A character that is used to denote items in a list. Common bullets include large dots, boxes, and diamonds.

**bulletin board system (BBS).** A telecommunications facility that enables users to share information with each other by using a modem and communications software.

**calligraphic.** A style of font that mimics handwriting.

**cap height.** The height of a face's capital letters.

**center justified.** Lines of type centered on the line length.

**character count.** The total number of characters in a document. Most word processing programs provide a character-count feature.

**characters per pica (CPP).** The number of characters per line pica. This number can be used to determine the length of copy. CPP should not be confused with pica typewriter type, which is 10 characters per inch.

**cicero.** European Didot system. A unit of measurement used mostly in Europe that relates roughly to the measurement of a pica. The point in the Didot system is larger than the U.S./British point. Twelve points in this system equal one Cicero.

**clipboard.** A temporary storage area that holds information that you cut or copied from a file.

**condensed type.** A narrower version of a typeface.

**copy fitting.** A method of calculating the length of text in a specific typeface and point size.

**CPP.** See *characters per pica.*

**crop.** To eliminate part of a photograph or drawing.

**crop marks.** The lines that appear on the outer edges of a page to mark the page size. The area outside the crop marks is trimmed off by the printer.

**descender.** The part of a letter that falls below the baseline. The downward strokes of the letters g, j, and p are examples of descenders.

**dingbats.** Characters that are symbols rather than letters. These characters include check marks, boxes, and arrows. Also known as pi characters.

**display face.** A typeface that is larger than 18 point. Display face is used in headlines.

**document.** A generic term that describes the work that you create by using a word processing or desktop publishing software package.

**dots per inch (DPI).** A measurement of the screen and printer resolution. The higher the number of dots, the better the resolution.

**downloadable font.** A font that is sent from your computer to your printer.

**drop cap.** A large capital letter that is used to start a paragraph or chapter. The letter is dropped into the surrounding text.

**Egyptian.** A style of type that is characterized by strong, thick serifs. This style originated in France after Napoleon's return from Egypt.

**ellipses.** The series of three periods that indicate an omission in text. Ellipses are most often used in quoted passages of text.

**em dash**. A long dash that is roughly equivalent to the length of a capital M. In desktop publishing, an em dash is used in place of two hyphens.

**em space.** A space that is equal to the width of a capital letter M in a specific typeface and point size.

**en dash.** A short dash that is roughly equivalent to the length of a capital N. An en dash is used in place of the word "to." An example is January–June.

**en space.** A unit of space that is the width of a capital letter N. An en space is half the width of an em space.

**EPS (encapsulated PostScript).** A file format that stores an image by using PostScript language.

**folio.** A folio is the page number of the document that appears at the bottom of the printed page.

**font.** Traditionally, the definition is the collection of letters, numbers, symbols, and punctuation marks that appear in the same typeface and size. With the advent of scalable font technology, a font is defined as the collection of letters, numbers, symbols, and punctuation marks that appear in the same typeface.

**font converter.** A program that converts a typeface from one format to another. You use font converters, for example, to convert TrueType fonts to Type 1 fonts.

**font editor.** A program that enables you to design your own typeface or to edit an existing typeface. Popular font editors include Fontographer and FontMonger.

**footer.** A line that runs along the bottom of your document. The line can contain a page number, the name of the document, or any information you want. See also *folio* and *header*.

**freeware.** Software that can be acquired without any cost to the user. Freeware can be distributed freely.

**gutter.** The inside white space of a page. Note that when you are printing a double-sided document, even numbered pages have an inside margin on the right side, odd-numbered pages on the left side.

**hairline.** A fine line or rule.

**hanging indent.** A type of indention in which the first line is set the full width of the line and following lines are indented a uniform amount.

**header.** A line that runs along the top of your document. The line can contain a page number, the name of the document, or any information that you want. See also *footer* and *folio*.

**hinting.** Instructions that scalable fonts use to make adjustments to the pixels making up a letter so that the letters look crisp and clear at lower resolutions. Hinting affects smaller point sizes. Points sizes above 14 or 18 points are not affected as much by hinting.

**hot-metal typesetting.** A form of typesetting in which the letters are cast in lead and then arranged in a form to be printed.

**italic.** A different cut of a typeface that appears right slanted. Italic was introduced by Aldus Manutius as a more economical, space-saving version of type. Italic type tends to be narrower with less letter spacing than the roman version of the same face. See also *oblique*.

**justified type.** Lines of text that line up on both the left and right margins. Letter spacing and word spacing are adjusted to even out the line.

**kerning.** Adjusting the space between certain letter pairs so that part of one character extends into the space of another character. Kerning is more prominent at larger point sizes. Common kerned pairs include WA, WO, PA, Tw, and Yo.

**landscape.** The orientation of a page in which the printed material runs horizontally across the longest side of the paper. See also *portrait*.

**leaders.** Rows of periods that help the reader's eye travel across a column. A table of contents usually lists the chapter name with leaders that run across to the page number.

**leading.** The amount of space between lines of type. You measure leading from baseline to baseline.

**letter spacing.** The spacing between letters. On computers, this term is known as *tracking*.

**ligature.** A character that is actually made up of two separate characters. Common ligatures are the letter combinations fi, fl, oe, and ae.

**line length.** The measurement of the horizontal width of a line of text. This measurement is also known as *line measure*.

**line spacing.** See *leading*.

**lowercase.** The small letters that make up the alphabet. The term comes from typographers storing the small letters in the lower part of a printer's typecase.

**modern.** A style of type in which the letters have extreme contrast between the thick and thin strokes with square serifs and strong vertical stress.

**monospace type.** A typeface in which all the characters are of the same width. See also *proportional type*.

**oblique.** A slanted version of a Roman typeface that simulates italic. See also *italic*.

**old style.** A type style characterized by small variations in the stroke weight of the letters, bracketed serifs, and diagonal stress.

**outline font.** A computerized font that can be scaled to any size before being printed as a bit map. Only one set of outlines is needed for each typeface.

**page description language (PDL).** A language used to describe printer output.

**page preview.** A feature of many word processing and desktop publishing programs that enables you to see the look of a page before it is printed.

**Pantone Matching System (PMS).** A standard system for matching specific colors. Designers and printers use this system to ensure that the printed document will contain the exact colors of choice.

**PCL 5.** The command language that the Hewlett-Packard LaserJet III printer uses. PCL 5 can also read bit-mapped fonts compatible with PCL 4.

**pica.** A standard unit of measurement in typography that is 12 points wide.

**pixel.** A single dot on your computer display. Pixels make up the characters that you see on-screen.

**point.** A standard unit of measurement in typography. One inch contains 72 points. One pica contains 12 points.

**portrait.** The orientation of a page in which the print runs along the shortest side of the paper. See also *landscape*.

**PostScript.** A page-description language written by Adobe, Inc. that prepares an image to be printed. PostScript fonts must be printed on PostScript-compatible printers. See also *EPS*.

**printer font.** A font that is designed for printing and not just for screen display.

**proofreader marks.** The shorthand type marks that proofreaders use to mark alterations and errors in copy.

**proportional spacing.** The characteristic in which wider letters, such as W, take up more horizontal space than narrower letters, such as i.

**public domain software.** Software that you can copy without copyright infringement. See also *freeware* and *shareware.*

**ragged right.** Lines of type that are printed so that the left margin is evenly aligned and the right margin is uneven.

**rasterize.** The conversion of the mathematical outlines of letters to create the filled-in character.

**resolution.** The number of dots per inch (dpi) that is displayed on-screen or on the printed document. The higher the dpi, the higher the resolution quality.

**reverse type.** The process of coloring the area behind a letter so that the characters appear white and the background appears dark. Reverse type is similar to a stencil sheet; the letters are the open or white area and the background is the dark or solid area. Reverse type is also known as knockout type.

**RIP.** Stands for raster image processor. A device that converts computer instructions into bit maps to be output by a printer.

**Roman.** A style of type characterized by upright letter strokes. On your word processor or desktop publishing program, Roman may be called Normal or Plain.

**run-around.** Type that fits around the shape of a picture of graphic.

**sans serif.** A typeface that contains no serifs. See also *serif.*

**screen font.** A bit-mapped font that is used to display type on your computer screen.

**serif.** A typeface that contains small cross strokes at the ends of letters. See also *sans serif.*

**shareware.** Computer programs that you can use on a trial basis. If you like the software, you are required to pay a fee to the program's developer.

**slab serif.** See *Egyptian.*

**small caps.** A style of type that makes uppercase letters smaller than full-size capitals.

**soft font.** A font that is stored on your computer, and then downloaded to your printer's RAM.

**solid leading.** Type that is set with no extra space between lines. An example of solid leading is 10 point type set on 10 points of leading.

**specing type.** Specifying the size, line length, and typeface of a document.

**stress.** The vertical, horizontal, or diagonal emphasis of a letter stroke.

**style.** A variation of a particular font. The variations can include boldface, italic, underline, shadow, and strikethrough.

**style sheet.** A collection of predefined text commands, such as type size, style, and alignment, that you can reuse to make formatting your documents easy.

**tracking.** The setting that determines the amount of letter spacing you use. Tracking can be set tight so that you eliminate space between letters, or loose so that you add more space between letters.

**transitional.** A subunit of Roman typefaces characterized by serifs that are less curved than old style faces and by character strokes that have more contrast between thick and thin.

**Type 1 fonts.** Adobe PostScript fonts that provide the highest output quality.

**Type 3 fonts.** A PostScript type format used by other font vendors before Adobe released Type 1 specifications. Usually, these fonts are not hinted.

**type manager.** A utility that organizes font definitions and renders font outlines on-screen and for a printer.

**type size.** The size of the letters that you use. Type size is measured in points.

**type style.** Variations in a typeface that include boldface, italic, underline, and so on.

**typeface.** A complete set of characters with similar characteristics.

**uncial.** A typeface that is characterized by calligraphic strokes. The name comes from the Latin word uncus, which means crooked.

**uppercase.** The capital letters that make up the alphabet. The term comes from typographers storing the capital letters in the upper part of a printer's typecase.

**weight.** The variation in a character's stroke width.

**white space.** The area in a design that contains no words or pictures. White space helps organize and separate the various elements of a page.

**word spacing.** The space between words in your document.

**x-height.** The height of lowercase letters. X-height does not include ascenders or descenders.

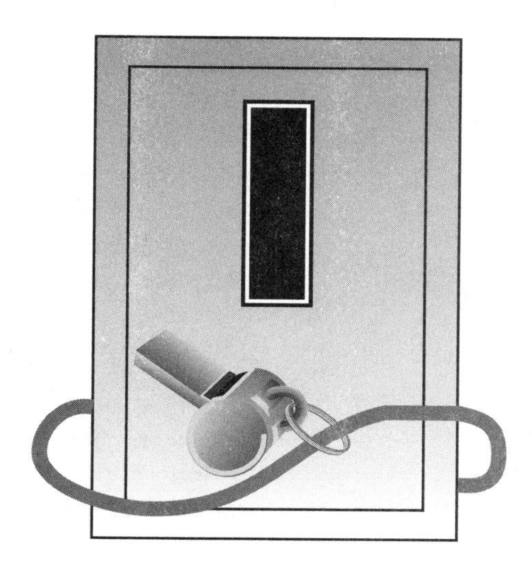

# INDEX

## SYMBOLS



## A

## D

## E



## F

## M



## N



## O



## P

## Q



## R

## T



## U



## V

## W

## X-Z